FREE Test Taking Tips DVD Offer

To help us bettRer serve you, we have developed a Test Taking Tips DVD that we would like to give you for FREE. **This DVD covers world-class test taking tips that you can use to be even more successful when you are taking your test.**

All that we ask is that you email us your feedback about your study guide. Please let us know what you thought about it – whether that is good, bad or indifferent.

To get your **FREE Test Taking Tips DVD**, email freedvd@studyguideteam.com with "FREE DVD" in the subject line and the following information in the body of the email:

    a. The title of your study guide.

    b. Your product rating on a scale of 1-5, with 5 being the highest rating.

    c. Your feedback about the study guide. What did you think of it?

    d. Your full name and shipping address to send your free DVD.

If you have any questions or concerns, please don't hesitate to contact us at freedvd@studyguideteam.com.

Thanks again!

Praxis Mathematics Content Knowledge 5161 Study Guide

Praxis II Math Content Knowledge Test 5161 Test Prep & Practice Test Questions

Test Prep Books Math Team

Table of Contents

Quick Overview 1

Test-Taking Strategies 2

FREE DVD OFFER 6

Introduction the Praxis II Mathematics Test 7

Number & Quantity, Algebra, Functions, and Calculus 9

    Number and Quantity 9

    Algebra 20

    Functions 36

    Calculus 51

    Practice Questions 61

    Answer Explanations 69

Geometry, Probability & Statistics, and Discrete Mathematics 75

    Geometry 75

    Probability and Statistics 93

    Discrete Mathematics 105

    Practice Questions 112

    Answer Explanations 116

Quick Overview

As you draw closer to taking your exam, effective preparation becomes more and more important. Thankfully, you have this study guide to help you get ready. Use this guide to help keep your studying on track and refer to it often.

This study guide contains several key sections that will help you be successful on your exam. The guide contains tips for what you should do the night before and the day of the test. Also included are test-taking tips. Knowing the right information is not always enough. Many well-prepared test takers struggle with exams. These tips will help equip you to accurately read, assess, and answer test questions.

A large part of the guide is devoted to showing you what content to expect on the exam and to helping you better understand that content. In this guide are practice test questions so that you can see how well you have grasped the content. Then, answer explanations are provided so that you can understand why you missed certain questions.

Don't try to cram the night before you take your exam. This is not a wise strategy for a few reasons. First, your retention of the information will be low. Your time would be better used by reviewing information you already know rather than trying to learn a lot of new information. Second, you will likely become stressed as you try to gain a large amount of knowledge in a short amount of time. Third, you will be depriving yourself of sleep. So be sure to go to bed at a reasonable time the night before. Being well-rested helps you focus and remain calm.

Be sure to eat a substantial breakfast the morning of the exam. If you are taking the exam in the afternoon, be sure to have a good lunch as well. Being hungry is distracting and can make it difficult to focus. You have hopefully spent lots of time preparing for the exam. Don't let an empty stomach get in the way of success!

When travelling to the testing center, leave earlier than needed. That way, you have a buffer in case you experience any delays. This will help you remain calm and will keep you from missing your appointment time at the testing center.

Be sure to pace yourself during the exam. Don't try to rush through the exam. There is no need to risk performing poorly on the exam just so you can leave the testing center early. Allow yourself to use all of the allotted time if needed.

Remain positive while taking the exam even if you feel like you are performing poorly. Thinking about the content you should have mastered will not help you perform better on the exam.

Once the exam is complete, take some time to relax. Even if you feel that you need to take the exam again, you will be well served by some down time before you begin studying again. It's often easier to convince yourself to study if you know that it will come with a reward!

Test-Taking Strategies

1. Predicting the Answer

When you feel confident in your preparation for a multiple-choice test, try predicting the answer before reading the answer choices. This is especially useful on questions that test objective factual knowledge. By predicting the answer before reading the available choices, you eliminate the possibility that you will be distracted or led astray by an incorrect answer choice. You will feel more confident in your selection if you read the question, predict the answer, and then find your prediction among the answer choices. After using this strategy, be sure to still read all of the answer choices carefully and completely. If you feel unprepared, you should not attempt to predict the answers. This would be a waste of time and an opportunity for your mind to wander in the wrong direction.

2. Reading the Whole Question

Too often, test takers scan a multiple-choice question, recognize a few familiar words, and immediately jump to the answer choices. Test authors are aware of this common impatience, and they will sometimes prey upon it. For instance, a test author might subtly turn the question into a negative, or he or she might redirect the focus of the question right at the end. The only way to avoid falling into these traps is to read the entirety of the question carefully before reading the answer choices.

3. Looking for Wrong Answers

Long and complicated multiple-choice questions can be intimidating. One way to simplify a difficult multiple-choice question is to eliminate all of the answer choices that are clearly wrong. In most sets of answers, there will be at least one selection that can be dismissed right away. If the test is administered on paper, the test taker could draw a line through it to indicate that it may be ignored; otherwise, the test taker will have to perform this operation mentally or on scratch paper. In either case, once the obviously incorrect answers have been eliminated, the remaining choices may be considered. Sometimes identifying the clearly wrong answers will give the test taker some information about the correct answer. For instance, if one of the remaining answer choices is a direct opposite of one of the eliminated answer choices, it may well be the correct answer. The opposite of obviously wrong is obviously right! Of course, this is not always the case. Some answers are obviously incorrect simply because they are irrelevant to the question being asked. Still, identifying and eliminating some incorrect answer choices is a good way to simplify a multiple-choice question.

4. Don't Overanalyze

Anxious test takers often overanalyze questions. When you are nervous, your brain will often run wild, causing you to make associations and discover clues that don't actually exist. If you feel that this may be a problem for you, do whatever you can to slow down during the test. Try taking a deep breath or counting to ten. As you read and consider the question, restrict yourself to the particular words used by the author. Avoid thought tangents about what the author *really* meant, or what he or she was *trying* to say. The only things that matter on a multiple-choice test are the words that are actually in the question. You must avoid reading too much into a multiple-choice question, or supposing that the writer meant something other than what he or she wrote.

5. No Need for Panic

It is wise to learn as many strategies as possible before taking a multiple-choice test, but it is likely that you will come across a few questions for which you simply don't know the answer. In this situation, avoid panicking. Because most multiple-choice tests include dozens of questions, the relative value of a single wrong answer is small. As much as possible, you should compartmentalize each question on a multiple-choice test. In other words, you should not allow your feelings about one question to affect your success on the others. When you find a question that you either don't understand or don't know how to answer, just take a deep breath and do your best. Read the entire question slowly and carefully. Try rephrasing the question a couple of different ways. Then, read all of the answer choices carefully. After eliminating obviously wrong answers, make a selection and move on to the next question.

6. Confusing Answer Choices

When working on a difficult multiple-choice question, there may be a tendency to focus on the answer choices that are the easiest to understand. Many people, whether consciously or not, gravitate to the answer choices that require the least concentration, knowledge, and memory. This is a mistake. When you come across an answer choice that is confusing, you should give it extra attention. A question might be confusing because you do not know the subject matter to which it refers. If this is the case, don't eliminate the answer before you have affirmatively settled on another. When you come across an answer choice of this type, set it aside as you look at the remaining choices. If you can confidently assert that one of the other choices is correct, you can leave the confusing answer aside. Otherwise, you will need to take a moment to try to better understand the confusing answer choice. Rephrasing is one way to tease out the sense of a confusing answer choice.

7. Your First Instinct

Many people struggle with multiple-choice tests because they overthink the questions. If you have studied sufficiently for the test, you should be prepared to trust your first instinct once you have carefully and completely read the question and all of the answer choices. There is a great deal of research suggesting that the mind can come to the correct conclusion very quickly once it has obtained all of the relevant information. At times, it may seem to you as if your intuition is working faster even than your reasoning mind. This may in fact be true. The knowledge you obtain while studying may be retrieved from your subconscious before you have a chance to work out the associations that support it. Verify your instinct by working out the reasons that it should be trusted.

8. Key Words

Many test takers struggle with multiple-choice questions because they have poor reading comprehension skills. Quickly reading and understanding a multiple-choice question requires a mixture of skill and experience. To help with this, try jotting down a few key words and phrases on a piece of scrap paper. Doing this concentrates the process of reading and forces the mind to weigh the relative importance of the question's parts. In selecting words and phrases to write down, the test taker thinks about the question more deeply and carefully. This is especially true for multiple-choice questions that are preceded by a long prompt.

9. Subtle Negatives

One of the oldest tricks in the multiple-choice test writer's book is to subtly reverse the meaning of a question with a word like *not* or *except*. If you are not paying attention to each word in the question, you can easily be led astray by this trick. For instance, a common question format is, "Which of the following is...?" Obviously, if the question instead is, "Which of the following is not...?," then the answer will be quite different. Even worse, the test makers are aware of the potential for this mistake and will include one answer choice that would be correct if the question were not negated or reversed. A test taker who misses the reversal will find what he or she believes to be a correct answer and will be so confident that he or she will fail to reread the question and discover the original error. The only way to avoid this is to practice a wide variety of multiple-choice questions and to pay close attention to each and every word.

10. Reading Every Answer Choice

It may seem obvious, but you should always read every one of the answer choices! Too many test takers fall into the habit of scanning the question and assuming that they understand the question because they recognize a few key words. From there, they pick the first answer choice that answers the question they believe they have read. Test takers who read all of the answer choices might discover that one of the latter answer choices is actually *more* correct. Moreover, reading all of the answer choices can remind you of facts related to the question that can help you arrive at the correct answer. Sometimes, a misstatement or incorrect detail in one of the latter answer choices will trigger your memory of the subject and will enable you to find the right answer. Failing to read all of the answer choices is like not reading all of the items on a restaurant menu: you might miss out on the perfect choice.

11. Spot the Hedges

One of the keys to success on multiple-choice tests is paying close attention to every word. This is never truer than with words like almost, most, some, and sometimes. These words are called "hedges" because they indicate that a statement is not totally true or not true in every place and time. An absolute statement will contain no hedges, but in many subjects, the answers are not always straightforward or absolute. There are always exceptions to the rules in these subjects. For this reason, you should favor those multiple-choice questions that contain hedging language. The presence of qualifying words indicates that the author is taking special care with his or her words, which is certainly important when composing the right answer. After all, there are many ways to be wrong, but there is only one way to be right! For this reason, it is wise to avoid answers that are absolute when taking a multiple-choice test. An absolute answer is one that says things are either all one way or all another. They often include words like *every*, *always*, *best*, and *never*. If you are taking a multiple-choice test in a subject that doesn't lend itself to absolute answers, be on your guard if you see any of these words.

12. Long Answers

In many subject areas, the answers are not simple. As already mentioned, the right answer often requires hedges. Another common feature of the answers to a complex or subjective question are qualifying clauses, which are groups of words that subtly modify the meaning of the sentence. If the question or answer choice describes a rule to which there are exceptions or the subject matter is complicated, ambiguous, or confusing, the correct answer will require many words in order to be expressed clearly and accurately. In essence, you should not be deterred by answer choices that seem excessively long. Oftentimes, the author of the text will not be able to write the correct answer without offering some qualifications and modifications. Your job is to read the answer choices thoroughly and completely and to select the one that most accurately and precisely answers the question.

13. Restating to Understand

Sometimes, a question on a multiple-choice test is difficult not because of what it asks but because of how it is written. If this is the case, restate the question or answer choice in different words. This process serves a couple of important purposes. First, it forces you to concentrate on the core of the question. In order to rephrase the question accurately, you have to understand it well. Rephrasing the question will concentrate your mind on the key words and ideas. Second, it will present the information to your mind in a fresh way. This process may trigger your memory and render some useful scrap of information picked up while studying.

14. True Statements

Sometimes an answer choice will be true in itself, but it does not answer the question. This is one of the main reasons why it is essential to read the question carefully and completely before proceeding to the answer choices. Too often, test takers skip ahead to the answer choices and look for true statements. Having found one of these, they are content to select it without reference to the question above. Obviously, this provides an easy way for test makers to play tricks. The savvy test taker will always read the entire question before turning to the answer choices. Then, having settled on a correct answer choice, he or she will refer to the original question and ensure that the selected answer is relevant. The mistake of choosing a correct-but-irrelevant answer choice is especially common on questions related to specific pieces of objective knowledge. A prepared test taker will have a wealth of factual knowledge at his or her disposal, and should not be careless in its application.

15. No Patterns

One of the more dangerous ideas that circulates about multiple-choice tests is that the correct answers tend to fall into patterns. These erroneous ideas range from a belief that B and C are the most common right answers, to the idea that an unprepared test-taker should answer "A-B-A-C-A-D-A-B-A." It cannot be emphasized enough that pattern-seeking of this type is exactly the WRONG way to approach a multiple-choice test. To begin with, it is highly unlikely that the test maker will plot the correct answers according to some predetermined pattern. The questions are scrambled and delivered in a random order. Furthermore, even if the test maker was following a pattern in the assignation of correct answers, there is no reason why the test taker would know which pattern he or she was using. Any attempt to discern a pattern in the answer choices is a waste of time and a distraction from the real work of taking the test. A test taker would be much better served by extra preparation before the test than by reliance on a pattern in the answers.

FREE DVD OFFER

Don't forget that doing well on your exam includes both understanding the test content and understanding how to use what you know to do well on the test. We offer a completely FREE Test Taking Tips DVD that covers world class test taking tips that you can use to be even more successful when you are taking your test.

All that we ask is that you email us your feedback about your study guide. To get your **FREE Test Taking Tips DVD**, email freedvd@studyguideteam.com with "FREE DVD" in the subject line and the following information in the body of the email:

- The title of your study guide.
- Your product rating on a scale of 1-5, with 5 being the highest rating.
- Your feedback about the study guide. What did you think of it?
- Your full name and shipping address to send your free DVD.

Introduction the Praxis II Mathematics Test

Function of the Praxis II Math Content Knowledge Test

The Praxis II Math Content Knowledge test is one of the Educational Testing Service's (ETS's) Subject Assessment tests that measures specific teaching skills and knowledge. The Praxis II Math Content Knowledge test is required by many states a part of the teacher licensing and certification process for those entering teaching in the field of Mathematics. Additionally, some professional organizations may require subject area testing for membership, even if subject-specific testing is not required for a teaching position.

Test Administration

All Praxis Subject Assessment Tests, including the Math Content Knowledge Test, are only administered in electronic format and in the English language. However, a version written in Braille is provided for those with certain visual impairments. Although there are not testing administration accommodations for other languages, extended testing time may be granted for candidates for whom English is a secondary language.

Depending upon the state requirements, tests may be offered continuously on a rolling basis throughout the year, or only administered during certain windows during the year. Prior to sitting for the exam, candidates must first register for the test online, submit the applicable fees, then report to the testing center.

Individuals with disabilities who would like test accommodations should register through the ETS Disability Services to have the requested accommodations approved. Online registration is not available for disability accommodations; applicants must make requests via mail or email. It takes approximately six weeks for the documentation to be reviewed. Accommodation request procedures are outlined in the *Bulletin Supplement for Test Takers with Disabilities or Health-related Needs*. Although online registration is not available, test scores may be viewed online, for which an account must be created.

Test Format

Registration for the Mathematics Content Knowledge Test comes with an interactive practice test that can be taken on the computer prior to the scored exam. As of September 1, 2016, calculators are provided for the Elementary Education version of the Mathematics Content Knowledge Test and a graphing calculator is permitted for specific sections only. The test is comprised of fill-in-the-blank and multiple-choice answer formats. Authorized testing time is generally set at 85 minutes, but can extend as far as two hours for single subject testing, when tests are taken separately.

Scores

Some Praxis tests offer the option to view unofficial test scores at the conclusion of the exam. When unofficial scores are not provided, it indicates that further analysis is required before the final scores can be rendered. Printed copies of the scores are not available at the test centers. On the Mathematics Content Knowledge Test, the selected response questions that are answered correctly are worth one point and scored by the computer. Constructed response questions are scored by educational professionals in mathematics and are scored by two different scorers. Scores can be verified through the

Praxis Score Information Bulletin. Below is a table that indicates the scoring breakdown for the Mathematics Content Knowledge:

- Possible Score Range: 100-200
- Score Interval: 1
- Number of Test Takers: 4,000
- Median Score: 165
- Average Score Range: 154-173

Number & Quantity, Algebra, Functions, and Calculus

Number and Quantity

Properties of Exponents

Exponents are used in mathematics to express a number or variable multiplied by itself a certain number of times. For example, x^3 means x is multiplied by itself three times. In this expression, x is called the **base**, and 3 is the **exponent**. Exponents can be used in more complex problems when they contain fractions and negative numbers.

Fractional exponents can be explained by looking first at the inverse of exponents, which are **roots.** Given the expression x^2, the square root can be taken, $\sqrt{x^2}$, cancelling out the 2 and leaving x by itself, if x is positive. Cancellation occurs because $\sqrt{x}$ can be written with exponents, instead of roots, as $x^{\frac{1}{2}}$. The numerator of 1 is the exponent, and the denominator of 2 is called the **root** (which is why it's referred to as a **square root**). Taking the square root of x^2 is the same as raising it to the $\frac{1}{2}$ power. Written out in mathematical form, it takes the following progression:

$$\sqrt{x^2} = (x^2)^{\frac{1}{2}} = x$$

From properties of exponents, $2 \times \frac{1}{2} = 1$ is the actual exponent of x. Another example can be seen with $x^{\frac{4}{7}}$. The variable x, raised to four-sevenths, is equal to the seventh root of x to the fourth power: $\sqrt[7]{x^4}$. In general,

$$x^{\frac{1}{n}} = \sqrt[n]{x}$$

and

$$x^{\frac{m}{n}} = \sqrt[n]{x^m}$$

Negative exponents also involve fractions. Whereas y^3 can also be rewritten as $\frac{y^3}{1}$, y^{-3} can be rewritten as $\frac{1}{y^3}$. A negative exponent means the exponential expression must be moved to the opposite spot in a fraction to make the exponent positive. If the negative appears in the numerator, it moves to the denominator. If the negative appears in the denominator, it is moved to the numerator. In general, $a^{-n} = \frac{1}{a^n}$, and a^{-n} and a^n are reciprocals.

Take, for example, the following expression:

$$\frac{a^{-4}b^2}{c^{-5}}$$

Since a is raised to the negative fourth power, it can be moved to the denominator. Since c is raised to the negative fifth power, it can be moved to the numerator. The b variable is raised to the positive second power, so it does not move.

The simplified expression is as follows:

$$\frac{b^2 c^5}{a^4}$$

In mathematical expressions containing exponents and other operations, the order of operations must be followed. **PEMDAS** states that exponents are calculated after any parenthesis and grouping symbols, but before any multiplication, division, addition, and subtraction.

Properties of Rational and Irrational Numbers

All real numbers can be separated into two groups: rational and irrational numbers. **Rational numbers** are any numbers that can be written as a fraction, such as $\frac{1}{3}, \frac{7}{4}$, and -25. Alternatively, **irrational numbers** are those that cannot be written as a fraction, such as numbers with never-ending, non-repeating decimal values. Many irrational numbers result from taking roots, such as $\sqrt{2}$ or $\sqrt{3}$. An irrational number may be written as:

$$34.5684952...$$

The **ellipsis** (...) represents the line of numbers after the decimal that does not repeat and is never-ending.

When rational and irrational numbers interact, there are different types of number outcomes. For example, when adding or multiplying two rational numbers, the result is a rational number. No matter what two fractions are added or multiplied together, the result can always be written as a fraction. The following expression shows two rational numbers multiplied together:

$$\frac{3}{8} \times \frac{4}{7} = \frac{12}{56}$$

The product of these two fractions is another fraction that can be simplified to $\frac{3}{14}$.

As another interaction, rational numbers added to irrational numbers will always result in irrational numbers. No part of any fraction can be added to a never-ending, non-repeating decimal to make a rational number. The same result is true when multiplying a rational and irrational number. Taking a fractional part of a never-ending, non-repeating decimal will always result in another never-ending, non-repeating decimal. An example of the product of rational and irrational numbers is shown in the following expression: $2 \times \sqrt{7}$.

The last type of interaction concerns two irrational numbers, where the sum or product may be rational or irrational depending on the numbers used. The following expression shows a rational sum from two irrational numbers:

$$\sqrt{3} + \left(6 - \sqrt{3}\right) = 6$$

The product of two irrational numbers can be rational or irrational. A rational result can be seen in the following expression:

$$\sqrt{2} \times \sqrt{8} = \sqrt{2 \times 8} = \sqrt{16} = 4$$

An irrational result can be seen in the following:

$$\sqrt{3} \times \sqrt{2} = \sqrt{6}$$

Solving Problems by Quantitative Reasoning

Dimensional analysis is the process of converting between different units using equivalent measurement statements. For instance, running 5 kilometers is approximately the same as running 3.1 miles. This conversion can be found by knowing that 1 kilometer is equal to approximately 0.62 miles.

When setting up the dimensional analysis calculations, the original units need to be opposite one another in each of the two fractions: one in the original amount (essentially in the numerator) and one in the denominator of the conversion factor. This enables them to cancel after multiplying, leaving the converted result.

Calculations involving formulas, such as determining volume and area, are a common situation in which units need to be interpreted and used. However, graphs can also carry meaning through units. The graph below is an example. It represents a graph of the position of an object over time. The y-axis represents the position or the number of meters the object is from the starting point at time s, in seconds. Interpreting this graph, the origin shows that at time zero seconds, the object is zero meters away from the starting point. As the time increases to one second, the position increases to five meters away.

This trend continues until 6 seconds, where the object is 30 meters away from the starting position. After this point in time—since the graph remains horizontal from 6 to 10 seconds—the object must have stopped moving.

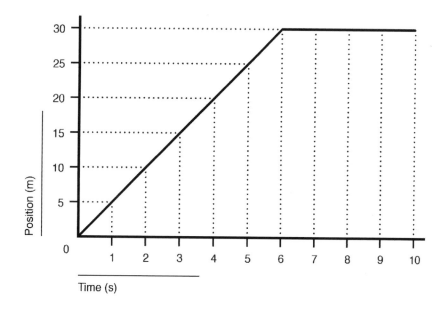

When solving problems with units, it's important to consider the reasonableness of the answer. If conversions are used, it's helpful to have an estimated value to compare the final answer to. This way, if the final answer is too distant from the estimate, it will be obvious that a mistake was made.

Structure of the Number System

The mathematical number system is made up of two general types of numbers: real and complex. **Real numbers** are those that are used in normal settings, while **complex numbers** are those composed of both a real number and an imaginary one. **Imaginary numbers** are the result of taking the square root of -1, and $\sqrt{-1} = i$.

The real number system is often explained using a Venn diagram similar to the one below. After a number has been labeled as a real number, further classification occurs when considering the other groups in this diagram. If a number is a never-ending, non-repeating decimal, it falls in the irrational category. Otherwise, it is rational. More information on these types of numbers is provided in the previous section. Furthermore, if a number does not have a fractional part, it is classified as an **integer,** such as -2, 75, or zero. **Whole numbers** are an even smaller group that only includes positive integers and zero. The last group of **natural numbers** is made up of only positive integers, such as 2, 56, or 12.

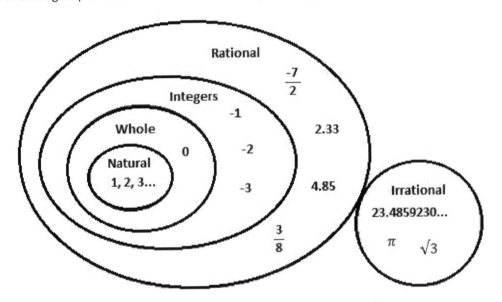

Real numbers can be compared and ordered using the number line. If a number falls to the left on the real number line, it is less than a number on the right. For example, $-2 < 5$ because -2 falls to the left of zero, and 5 falls to the right. Numbers to the left of zero are negative while those to the right are positive.

Complex numbers are made up of the sum of a real number and an imaginary number. Some examples of complex numbers include $6 + 2i$, $5 - 7i$, and $-3 + 12i$. Adding and subtracting complex numbers is similar to collecting like terms. The real numbers are added together, and the imaginary numbers are added together. For example, if the problem asks to simplify the expression $6 + 2i - 3 + 7i$, the 6 and (-3) are combined to make 3, and the $2i$ and $7i$ combine to make $9i$. Multiplying and dividing complex numbers is similar to working with exponents. One rule to remember when multiplying is that $i \times i = -1$. For example, if a problem asks to simplify the expression $4i(3 + 7i)$, the $4i$ should be distributed throughout the 3 and the $7i$. This leaves the final expression $12i - 28$. The 28 is negative because $i \times i$ results in a negative number. The last type of operation to consider with complex numbers is the conjugate. The **conjugate** of a complex number is a technique used to change the complex number into a real number. For example, the conjugate of $4 - 3i$ is $4 + 3i$. Multiplying $(4 - 3i)(4 + 3i)$ results in $16 + 12i - 12i + 9$, which has a final answer of $16 + 9 = 25$.

The order of operations—PEMDAS—simplifies longer expressions with real or imaginary numbers. Each operation is listed in the order of how they should be completed in a problem containing more than one operation. Parenthesis can also mean grouping symbols, such as brackets and absolute value. Then, exponents are calculated. Multiplication and division should be completed from left to right, and addition and subtraction should be completed from left to right.

Simplification of another type of expression occurs when radicals are involved. As explained previously, *root* is another word for *radical*. For example, the following expression is a radical that can be simplified: $\sqrt{24x^2}$. First, the number must be factored out to the highest perfect square. Any perfect square can be taken out of a radical. Twenty-four can be factored into 4 and 6, and 4 can be taken out of the radical. $\sqrt{4} = 2$ can be taken out, and 6 stays underneath. If $x > 0$, x can be taken out of the radical because it is a perfect square. The simplified radical is $2x\sqrt{6}$. An approximation can be found using a calculator.

There are also properties of numbers that are true for certain operations. The **commutative property** allows the order of the terms in an expression to change while keeping the same final answer. Both addition and multiplication can be completed in any order and still obtain the same result. However, order does matter in subtraction and division. The **associative property** allows any terms to be "associated" by parenthesis and retain the same final answer. For example, $(4 + 3) + 5 = 4 + (3 + 5)$. Both addition and multiplication are associative; however, subtraction and division do not hold this property. The **distributive property** states that $a(b + c) = ab + ac$. It is a property that involves both addition and multiplication, and the a is distributed onto each term inside the parentheses.

Integers can be factored into prime numbers. To **factor** is to express as a product. For example, $6 = 3 \times 2$, and $6 = 6 \times 1$. Both are factorizations, but the expression involving the factors of 3 and 2 is known as a **prime factorization** because it is factored into a product of two **prime numbers**—integers that do not have any factors other than themselves and 1. A **composite number** is a positive integer that can be divided into at least one other integer other than itself and 1, such as 6. Integers that have a factor of 2 are even, and if they are not divisible by 2, they are odd. Finally, a **multiple** of a number is the product of that number and a **counting number**—also known as a **natural number**. For example, some multiples of 4 are 4, 8, 12, 16, etc.

Complex Numbers as Solutions

Complex numbers may result from solving polynomial equations using the quadratic equation. Since complex numbers result from taking the square root of a negative number, the number found under the radical in the quadratic formula—called the **determinant**—determines whether or not the answer will be real or complex. If the determinant is negative, the roots are complex. Even though the coefficients of the polynomial may be real numbers, the roots are complex.

Solving polynomials by factoring is an alternative to using the quadratic formula. For example, in order to solve $x^2 - b^2 = 0$ for x, it needs to be factored. It factors into $(x + b)(x - b) = 0$. The solution set can be found by setting each factor equal to zero, resulting in $x = \pm b$. When b^2 is negative, the factors are complex numbers. For example, $x^2 + 64 = 0$ can be factored into $(x + 8i)(x - 8i) = 0$. The two roots are then found to be $x = \pm 8i$.

When dealing with polynomials and solving polynomial equations, it is important to remember the **fundamental theorem of algebra**. When given a polynomial with a degree of n, the theorem states that there will be n roots. These roots may or may not be complex. For example, the following polynomial equation of degree 2 has two complex roots: $x^2 + 1 = 0$. The factors of this polynomial are $(x + i)$ and

$(x - i)$, resulting in the roots $x = i, -i$. As seen on the graph below, imaginary roots occur when the graph does not touch the x-axis.

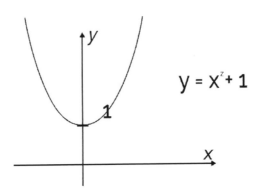

When a graphing calculator is permitted, the graph can always confirm the number and types of roots of the polynomial.

A **polynomial identity** is a true equation involving polynomials. For example:

$$x^2 - 5x + 6 = (x - 3)(x - 2)$$

This can be proved through multiplication by the FOIL method and factoring. This idea can be extended to involve complex numbers. For example:

$$i^2 = -1, x^3 + 9x = x(x^2 + 9) = x(x + \sqrt{3}i)(x - \sqrt{3}i)$$

This identity can also be proven through FOIL and factoring.

Recall that the FOIL method is a strategy used to remember how to multiply two binomials. FOIL stands for "First, Outer, Inner, Last" because the product of two binomials is the sum of multiplying the *first* terms, multiplying the *outer* terms, multiplying the *inner* terms, and multiplying the *last* terms. For example, the product of $(a + b) \times (c + d) = ac + ad + bc + bd$.

Matrices

Matrices can be used to represent linear equations, solve systems of equations, and manipulate data to simulate change. Matrices consist of numerical entries in both rows and columns. The following matrix A is a 3 x 4 matrix because it has three rows and four columns:

$$A = \begin{bmatrix} 3 & 2 & -5 & 3 \\ 3 & 6 & 2 & -5 \\ -1 & 3 & 7 & 0 \end{bmatrix}$$

Matrices can be added or subtracted only if they have the same dimensions. For example, the following matrices can be added by adding corresponding matrix entries:

$$\begin{bmatrix} 3 & 4 \\ 2 & -6 \end{bmatrix} + \begin{bmatrix} -1 & 4 \\ 4 & 2 \end{bmatrix} = \begin{bmatrix} 2 & 8 \\ 6 & -4 \end{bmatrix}$$

Multiplication can also be used to manipulate matrices. **Scalar multiplication** involves multiplying a matrix by a constant. Each matrix entry needs to be multiplied by the constant. The following example shows a 3×2 matrix being multiplied by the constant 6:

$$6 \times \begin{bmatrix} 3 & 4 \\ 2 & -6 \\ 1 & 0 \end{bmatrix} = \begin{bmatrix} 18 & 24 \\ 12 & -36 \\ 6 & 0 \end{bmatrix}$$

Matrix multiplication of two matrices involves finding multiple dot products. The **dot product** of a row and column is the sum of the products of each corresponding row and column entry. In the following example, a 2×2 matrix is multiplied by a 2×2 matrix. The dot product of the first row and column is:

$$(1 \times 2) + (2 \times 1) = (2) + (2) = 4$$

$$\begin{bmatrix} 1 & 2 \\ 3 & 4 \end{bmatrix} \times \begin{bmatrix} 2 & 0 \\ 1 & 2 \end{bmatrix} = \begin{bmatrix} 4 & 4 \\ 10 & 8 \end{bmatrix}$$

The same process is followed to find the other three values in the solution matrix. Matrices can only be multiplied if the number of columns in the first matrix equals the number of rows in the second matrix. The previous example is also an example of square matrix multiplication because they are both square matrices. A **square matrix** has the same number of rows and columns. For square matrices, the order in which they are multiplied does matter. Therefore, matrix multiplication does not satisfy the commutative property. It does, however, satisfy the associative and distributive properties.

Another transformation of matrices can be found by using the **identity matrix**—also referred to as the "I" **matrix**. The identity matrix is similar to the number one in normal multiplication. The identity matrix is a square matrix with ones in the diagonal spots and zeros everywhere else. The identity matrix is also the result of multiplying a matrix by its inverse. This process is similar to multiplying a number by its reciprocal.

The **zero matrix** is also a matrix acting as an additive identity. The zero matrix consists of zeros in every entry. It does not change the values of a matrix when using addition.

Given a system of linear equations, a matrix can be used to represent the entire system. Operations can then be performed on the matrix to solve the system. The following system offers an example:

$$\begin{aligned} x + y + z &= 6 \\ y + 5z &= -4 \\ 2x + 5y - z &= 27 \end{aligned}$$

There are three variables and three equations. The coefficients in the equations can be used to form a 3 x 3 matrix:

$$\begin{bmatrix} 1 & 1 & 1 \\ 0 & 1 & 5 \\ 2 & 5 & -1 \end{bmatrix}$$

The number of rows equals the number of equations, and the number of columns equals the number of variables. The numbers on the right side of the equations can be turned into a 3 x 1 matrix. That matrix is shown here:

$$\begin{bmatrix} 6 \\ -4 \\ 27 \end{bmatrix}$$

It can also be referred to as a **vector**. The variables are represented in a matrix of their own:

$$\begin{bmatrix} x \\ y \\ z \end{bmatrix}$$

The system can be represented by the following matrix equation:

$$\begin{bmatrix} 1 & 1 & 1 \\ 0 & 2 & 5 \\ 2 & 5 & -1 \end{bmatrix} \begin{bmatrix} x \\ y \\ z \end{bmatrix} = \begin{bmatrix} 6 \\ -4 \\ 27 \end{bmatrix}$$

Simply, this is written as $AX = B$. By using the inverse of a matrix, the solution can be found: $X = A^{-1}B$. Once the inverse of A is found using operations, it is then multiplied by B to find the solution to the system: $x = 5, y = 3,$ and $z = -2$.

The determinant of a 2 x 2 matrix is the following:

$$|A| = \begin{vmatrix} a & b \\ c & d \end{vmatrix} = ad - bc$$

It is a number related to the size of the matrix. The absolute value of the determinant of matrix A is equal to the area of a parallelogram with vertices (0, 0), (a, b), (c, d), and (a+b, c+d).

Ratios and Proportions

Ratios are used to show the relationship between two quantities. The ratio of oranges to apples in the grocery store may be 3 to 2. That means that for every 3 oranges, there are 2 apples. This comparison can be expanded to represent the actual number of oranges and apples. Another example may be the number of boys to girls in a math class. If the ratio of boys to girls is given as 2 to 5, that means there are 2 boys to every 5 girls in the class. Ratios can also be compared if the units in each ratio are the same. The ratio of boys to girls in the math class can be compared to the ratio of boys to girls in a science class by stating which ratio is higher and which is lower.

Rates are used to compare two quantities with different units. **Unit rates** are the simplest form of rate. With unit rates, the denominator in the comparison of two units is one. For example, if someone can type at a rate of 1000 words in 5 minutes, then his or her unit rate for typing is $\frac{1000}{5} = 200$ words in one minute or 200 words per minute. Any rate can be converted into a unit rate by dividing to make the denominator one. 1000 words in 5 minutes has been converted into the unit rate of 200 words per minute.

Ratios and rates can be used together to convert rates into different units. For example, if someone is driving 50 kilometers per hour, that rate can be converted into miles per hour by using a ratio known as the **conversion factor**. Since the given value contains kilometers and the final answer needs to be in miles, the ratio relating miles to kilometers needs to be used. There are 0.62 miles in 1 kilometer. This, written as a ratio and in fraction form, is:

$$\frac{0.62 \; miles}{1 \; km}$$

To convert 50km/hour into miles per hour, the following conversion needs to be set up:

$$\frac{50 \; km}{hour} \times \frac{0.62 \; miles}{1 \; km} = 31 \; miles \; per \; hour$$

The ratio between two similar geometric figures is called the **scale factor**. For example, a problem may depict two similar triangles, A and B. The scale factor from the smaller triangle A to the larger triangle B is given as 2 because the length of the corresponding side of the larger triangle, 16, is twice the corresponding side on the smaller triangle, 8. This scale factor can also be used to find the value of a missing side, x, in triangle A. Since the scale factor from the smaller triangle (A) to larger one (B) is 2, the larger corresponding side in triangle B (given as 25), can be divided by 2 to find the missing side in A ($x =$ 12.5). The scale factor can also be represented in the equation $2A = B$ because two times the lengths of A gives the corresponding lengths of B. This is the idea behind similar triangles.

Much like a scale factor can be written using an equation like $2A = B$, a **relationship** is represented by the equation $Y = kX$. X and Y are proportional because as values of X increase, the values of Y also increase. A relationship that is inversely proportional can be represented by the equation $Y = \frac{k}{x}$, where the value of Y decreases as the value of x increases and vice versa.

Proportional reasoning can be used to solve problems involving ratios, percentages, and averages. Ratios can be used in setting up proportions and solving them to find unknowns. For example, if a student completes an average of 10 pages of math homework in 3 nights, how long would it take the student to complete 22 pages? Both ratios can be written as fractions. The second ratio would contain the unknown.

The following proportion represents this problem, where x is the unknown number of nights:

$$\frac{10 \; pages}{3 \; nights} = \frac{22 \; pages}{x \; nights}$$

Solving this proportion entails cross-multiplying and results in the following equation: $10x = 22 \times 3$. Simplifying and solving for x results in the exact solution: $x = 6.6 \; nights$. The result would be rounded up to 7 because the homework would actually be completed on the 7th night.

The following problem uses ratios involving percentages:

If 20% of the class is girls and 30 students are in the class, how many girls are in the class?

To set up this problem, it is helpful to use the common proportion:

$$\frac{\%}{100} = \frac{is}{of}$$

Within the proportion, % is the percentage of girls, 100 is the total percentage of the class, *is* is the number of girls, and *of* is the total number of students in the class. Most percentage problems can be written using this language. To solve this problem, the proportion should be set up as $\frac{20}{100} = \frac{x}{30}$, and then solved for x. Cross-multiplying results in the equation $20 \times 30 = 100x$, which results in the solution $x = 6$. There are 6 girls in the class.

Problems involving volume, length, and other units can also be solved using ratios. A problem may ask for the volume of a cone to be found that has a radius, $r = 7m$ and a height, $h = 16m$. Referring to the formulas provided on the test, the volume of a cone is given as:

$$V = \pi r^2 \frac{h}{3}$$

r is the radius, and h is the height. Plugging $r = 7$ and $h = 16$ into the formula, the following is obtained:

$$V = \pi (7^2) \frac{16}{3}$$

Therefore, volume of the cone is found to be approximately 821m³. Sometimes, answers in different units are sought. If this problem wanted the answer in liters, 821m³ would need to be converted.

Using the equivalence statement 1m³ = 1000L, the following ratio would be used to solve for liters:

$$821m^3 \times \frac{1000L}{1m^3}$$

Cubic meters in the numerator and denominator cancel each other out, and the answer is converted to 821,000 liters, or 8.21×10^5 L.

Other conversions can also be made between different given and final units. If the temperature in a pool is 30°C, what is the temperature of the pool in degrees Fahrenheit? To convert these units, an equation is used relating Celsius to Fahrenheit. The following equation is used:

$$T_{°F} = 1.8 T_{°C} + 32$$

Plugging in the given temperature and solving the equation for T yields the result:

$$T_{°F} = 1.8(30) + 32 = 86°F$$

Both units in the metric system and U.S. customary system are widely used.

Precision and Accuracy

Precision and accuracy are used to describe groups of measurements. **Precision** describes a group of measures that are very close together, regardless of whether the measures are close to the true value. **Accuracy** describes how close the measures are to the true value.

Since accuracy refers to the closeness of a value to the true measurement, the level of accuracy depends on the object measured and the instrument used to measure it. This will vary depending on the situation. If measuring the mass of a set of dictionaries, kilograms may be used as the units. In this case, it is not vitally important to have a high level of accuracy. If the measurement is a few grams away from the true value, the discrepancy might not make a big difference in the problem.

In a different situation, the level of accuracy may be more significant. Pharmacists need to be sure they are very accurate in their measurements of medicines that they give to patients. In this case, the level of accuracy is vitally important and not something to be estimated. In the dictionary situation, the measurements were given as whole numbers in kilograms. In the pharmacist's situation, the measurements for medicine must be taken to the milligram and sometimes further, depending on the type of medicine.

When considering the accuracy of measurements, the error in each measurement can be shown as absolute and relative. **Absolute error** tells the actual difference between the measured value and the true value. The **relative error** tells how large the error is in relation to the true value. There may be two problems where the absolute error of the measurements is 10 grams. For one problem, this may mean the relative error is very small because the measured value is 14,990 grams, and the true value is 15,000 grams. Ten grams in relation to the true value of 15,000 is small: 0.06%. For the other problem, the measured value is 290 grams, and the true value is 300 grams. In this case, the 10-gram absolute error means a high relative error because the true value is smaller. The relative error is 10/300 = 0.03, or 3%.

Scientific Notation

Scientific Notation is used to represent numbers that are either very small or very large. For example, the distance to the Sun is approximately 150,000,000,000 meters. Instead of writing this number with so many zeros, it can be written in scientific notation as 1.5×10^{11} meters. The same is true for very small numbers, but the exponent becomes negative. If the mass of a human cell is 0.000000000001 kilograms, that measurement can be easily represented by 1.0×10^{-12} kilograms. In both situations, scientific notation makes the measurement easier to read and understand. Each number is translated to an expression with one digit in the tens place multiplied by an expression corresponding to the zeros.

When two measurements are given and both involve scientific notation, it is important to know how these interact with each other:

- In addition and subtraction, the exponent on the ten must be the same before any operations are performed on the numbers. For example, $(1.3 \times 10^4) + (3.0 \times 10^3)$ cannot be added until one of the exponents on the ten is changed. The 3.0×10^3 can be changed to 0.3×10^4, then the 1.3 and 0.3 can be added. The answer comes out to be 1.6×10^4.

- For multiplication, the first numbers can be multiplied and then the exponents on the tens can be added. Once an answer is formed, it may have to be converted into scientific notation again depending on the change that occurred.

- The following is an example of multiplication with scientific notation:

$$(4.5 \times 10^3) \times (3.0 \times 10^{-5}) = 13.5 \times 10^{-2}$$

- Since this answer is not in scientific notation, the decimal is moved over to the left one unit, and 1 is added to the ten's exponent. This results in the final answer: 1.35×10^{-1}.

- For division, the first numbers are divided, and the exponents on the tens are subtracted. Again, the answer may need to be converted into scientific notation form, depending on the type of changes that occurred during the problem.

- **Order of magnitude** relates to scientific notation and is the total count of powers of 10 in a number. For example, there are 6 orders of magnitude in 1,000,000. If a number is raised by an order of magnitude, it is multiplied by 10. Order of magnitude can be helpful in estimating results using very large or small numbers. An answer should make sense in terms of its order of magnitude.

- For example, if area is calculated using two dimensions with 6 orders of magnitude, because area involves multiplication, the answer should have around 12 orders of magnitude. Also, answers can be estimated by rounding to the largest place value in each number. For example, 5,493,302 ×2,523,100 can be estimated by 5 × 3 = 15 with 6 orders of magnitude.

Algebra

Rewriting Expressions

Algebraic expressions are made up of numbers, variables, and combinations of the two, using mathematical operations. Expressions can be rewritten based on their factors. For example, the expression $6x + 4$ can be rewritten as $2(3x + 2)$ because 2 is a factor of both $6x$ and 4. More complex expressions can also be rewritten based on their factors. The expression $x^4 - 16$ can be rewritten as $(x^2 - 4)(x^2 + 4)$. This is a different type of factoring, where a difference of squares is factored into a sum and difference of the same two terms. With some expressions, the factoring process is simple and only leads to a different way to represent the expression. With others, factoring and rewriting the expression leads to more information about the given problem.

In the following quadratic equation, factoring the binomial leads to finding the zeros of the function:

$$x^2 - 5x + 6 = y$$

This equations factors into $(x - 3)(x - 2) = y$, where 2 and 3 are found to be the zeros of the function when y is set equal to zero. The zeros of any function are the x-values where the graph of the function on the coordinate plane crosses the x-axis.

Factoring an equation is a simple way to rewrite the equation and find the zeros, but factoring is not possible for every quadratic. Completing the square is one way to find zeros when factoring is not an option. The following equation cannot be factored: $x^2 + 10x - 9 = 0$. The first step in this method is to move the constant to the right side of the equation, making it $x^2 + 10x = 9$. Then, the coefficient of x is divided by 2 and squared. This number is then added to both sides of the equation, to make the equation still true. For this example, $\left(\frac{10}{2}\right)^2 = 25$ is added to both sides of the equation to obtain:

$$x^2 + 10x + 25 = 9 + 25$$

This expression simplifies to $x^2 + 10x + 25 = 34$, which can then be factored into $(x + 5)^2 = 34$. Solving for x then involves taking the square root of both sides and subtracting 5.

This leads to two zeros of the function:

$$x = \pm\sqrt{34} - 5$$

Depending on the type of answer the question seeks, a calculator may be used to find exact numbers.

Given a **quadratic equation in standard form**— $ax^2 + bx + c = 0$—the sign of a tells whether the function has a minimum value or a maximum value. If $a > 0$, the graph opens up and has a minimum value. If $a < 0$, the graph opens down and has a maximum value. Depending on the way the quadratic equation is written, multiplication may need to occur before a max/min value is determined.

Exponential expressions can also be rewritten, just as quadratic equations. Properties of exponents must be understood. Multiplying two exponential expressions with the same base involves adding the exponents:

$$a^m a^n = a^{m+n}$$

Dividing two exponential expressions with the same base involves subtracting the exponents:

$$\frac{a^m}{a^n} = a^{m-n}$$

Raising an exponential expression to another exponent includes multiplying the exponents:

$$(a^m)^n = a^{mn}$$

The zero power always gives a value of 1: $a^0 = 1$. Raising either a product or a fraction to a power involves distributing that power:

$$(ab)^m = a^m b^m \text{ and } \left(\frac{a}{b}\right)^m = \frac{a^m}{b^m}$$

Finally, raising a number to a negative exponent is equivalent to the reciprocal including the positive exponent:

$$a^{-m} = \frac{1}{a^m}$$

Operations with Polynomials

Addition and subtraction operations can be performed on polynomials with like terms. **Like terms** refers to terms that have the same variable and exponent. The two following polynomials can be added together by collecting like terms:

$$(x^2 + 3x - 4) + (4x^2 - 7x + 8)$$

The x^2 terms can be added as $x^2 + 4x^2 = 5x^2$. The x terms can be added as $3x + -7x = -4x$, and the constants can be added as $-4 + 8 = 4$. The following expression is the result of the addition:

$$5x^2 - 4x + 4$$

When subtracting polynomials, the same steps are followed, only subtracting like terms together.

Multiplication of polynomials can also be performed. Given the two polynomials, $(y^3 - 4)$ and $(x^2 + 8x - 7)$, each term in the first polynomial must be multiplied by each term in the second polynomial. The steps to multiply each term in the given example are as follows:

$$(y^3 \times x^2) + (y^3 \times 8x) + (y^3 \times -7) + (-4 \times x^2) + (-4 \times 8x) + (-4 \times -7)$$

Simplifying each multiplied part, yields:

$$x^2y^3 + 8xy^3 - 7y^3 - 4x^2 - 32x + 28$$

None of the terms can be combined because there are no like terms in the final expression. Any polynomials can be multiplied by each other by following the same set of steps, then collecting like terms at the end.

Zeros of Polynomials

Finding the **zeros of polynomial functions** is the same process as finding the solutions of polynomial equations. These are the points at which the graph of the function crosses the x-axis. As stated previously, factors can be used to find the zeros of a polynomial function. The degree of the function shows the number of possible zeros. If the highest exponent on the independent variable is 4, then the degree is 4, and the number of possible zeros is 4. If there are complex solutions, the number of roots is less than the degree.

Given the function $y = x^2 + 7x + 6$, y can be set equal to zero, and the polynomial can be factored. The equation turns into $0 = (x + 1)(x + 6)$, where $x = -1$ and $x = -6$ are the zeros. Since this is a quadratic equation, the shape of the graph will be a parabola. Knowing that zeros represent the points where the parabola crosses the x-axis, the maximum or minimum point is the only other piece needed to sketch a rough graph of the function. By looking at the function in standard form, the coefficient of x is positive; therefore, the parabola opens up. Using the zeros and the minimum, the following rough sketch of the graph can be constructed:

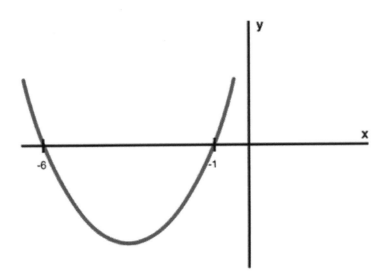

Polynomial Identities

Difference of squares refers to a binomial composed of the difference of two squares. For example, $a^2 - b^2$ is a difference of squares. It can be written $(a)^2 - (b)^2$, and it can be factored into $(a - b)(a + b)$. Recognizing the difference of squares allows the expression to be rewritten easily because of the form it takes. For some expressions, factoring consists of more than one step. When factoring, it's important to always check to make sure that the result cannot be factored further. If it can, then the expression should be split further. If it cannot be, the factoring step is complete, and the expression is completely factored.

A sum and difference of cubes is another way to factor a polynomial expression. When the polynomial takes the form of addition or subtraction of two terms that can be written as a cube, a formula is given. The following graphic shows the factorization of a **difference of cubes**:

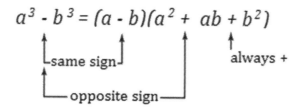

This form of factoring can be useful in finding the zeros of a function of degree 3. For example, when solving $x^3 - 27 = 0$, this rule needs to be used. $x^3 - 27$ is first written as the difference two cubes, $(x)^3 - (3)^3$ and then factored into $(x - 3)(x^2 + 3x + 9)$. This expression may not be factored any further. Each factor is then set equal to zero. Therefore, one solution is found to be $x = 3$, and the other two solutions must be found using the quadratic formula. A sum of squares would have a similar process. The formula for factoring a **sum of cubes** is:

$$a^3 + b^3 = (a + b)(a^2 - ab + b^2)$$

The opposite of factoring is multiplying. Multiplying a square of a binomial involves the following rules:

$$(a + b)^2 = a^2 + 2ab + b^2$$

$$(a - b)^2 = a^2 - 2ab + b^2$$

The **binomial theorem** for expansion can be used when the exponent on a binomial is larger than 2, and the multiplication would take a long time. The binomial theorem is given as:

$$(a + b)^n = \sum_{k=0}^{n} \binom{n}{k} a^{n-k} b^k$$

$$\text{where} \quad \binom{n}{k} = \frac{n!}{k!(n-k)!}$$

For example, $(a + b)^5$ would become $a^5 + 5a^4 b + 10a^3 b^2 + 10a^2 b^3 + 5ab^4 + b^5$.

The **Remainder Theorem** can be helpful when evaluating polynomial functions $P(x)$ for a given value of x. A polynomial can be divided by $(x - a)$, if there is a remainder of 0. This also means that $P(a) = 0$ and $(x - a)$ is a factor of $P(x)$. In a similar sense, if P is evaluated at any other number b, $P(b)$ is equal to the remainder of dividing $P(x)$ by $(x - b)$.

Rational Expressions

A fraction, or ratio, wherein each part is a polynomial, defines **rational expressions**. Some examples include $\frac{2x+6}{x}$, $\frac{1}{x^2-4x+8}$, and $\frac{z^2}{x+5}$. Exponents on the variables are restricted to whole numbers, which means roots and negative exponents are not included in rational expressions.

Rational expressions can be transformed by factoring. For example, the expression $\frac{x^2-5x+6}{(x-3)}$ can be rewritten by factoring the numerator to obtain $\frac{(x-3)(x-2)}{(x-3)}$. Therefore, the common binomial $(x - 3)$ can cancel so that the simplified expression is $\frac{(x-2)}{1} = (x - 2)$.

Additionally, other rational expressions can be rewritten to take on different forms. Some may be factorable in themselves, while others can be transformed through arithmetic operations. Rational expressions are closed under addition, subtraction, multiplication, and division by a nonzero expression. **Closed** means that if any one of these operations is performed on a rational expression, the result will still be a rational expression. The set of all real numbers is another example of a set closed under all four operations.

Adding and subtracting rational expressions is based on the same concepts as adding and subtracting simple fractions. For both concepts, the denominators must be the same for the operation to take place. For example, here are two rational expressions:

$$\frac{x^3 - 4}{(x - 3)} + \frac{x + 8}{(x - 3)}$$

Since the denominators are both $(x - 3)$, the numerators can be combined by collecting like terms to form:

$$\frac{x^3 + x + 4}{(x - 3)}$$

If the denominators are different, they need to be made common (the same) by using the **Least Common Denominator (LCD)**. Each denominator needs to be factored, and the LCD contains each factor that appears in any one denominator the greatest number of times it appears in any denominator. The original expressions need to be multiplied by a form of 1 such as $\frac{5}{5}$ or $\frac{x-2}{x-2}$, which will turn each denominator into the LCD. This process is like adding fractions with unlike denominators. It is also important when working with rational expressions to define what value of the variable makes the denominator zero. For this particular value, the expression is undefined.

Multiplication of rational expressions is performed like multiplication of fractions. The numerators are multiplied; then, the denominators are multiplied. The final fraction is then simplified. The expressions are simplified by factoring and cancelling out common terms. In the following example, the numerator of the second expression can be factored first to simplify the expression before multiplying:

$$\frac{x^2}{(x - 4)} \times \frac{x^2 - x - 12}{2}$$

$$\frac{x^2}{(x - 4)} \times \frac{(x - 4)(x + 3)}{2}$$

The $(x - 4)$ on the top and bottom cancel out:

$$\frac{x^2}{1} \times \frac{(x + 3)}{2}$$

Then multiplication is performed, resulting in:

$$\frac{x^3 + 3x^2}{2}$$

Dividing rational expressions is similar to the division of fractions, where division turns into multiplying by a reciprocal. Thus, the following expression can be rewritten as a multiplication problem:

$$\frac{x^2 - 3x + 7}{x - 4} \div \frac{x^2 - 5x + 3}{x - 4}$$

$$\frac{x^2 - 3x + 7}{x - 4} \times \frac{x - 4}{x^2 - 5x + 3}$$

The $x - 4$ cancels out, leaving:

$$\frac{x^2 - 3x + 7}{x^2 - 5x + 3}$$

The final answers should always be completely simplified. If a function is composed of a rational expression, the zeros of the graph can be found from setting the polynomial in the numerator as equal to zero and solving. The values that make the denominator equal to zero will either exist on the graph as a **hole** or a **vertical asymptote**.

Equations and Inequalities

Imagine the following problem: The sum of a number and 5 is equal to -8 times the number.

To find this unknown number, a simple equation can be written to represent the problem. Key words such as difference, equal, and times are used to form the following equation with one variable: $n + 5 = -8n$. When solving for n, opposite operations are used. First, n is subtracted from $-8n$ across the equals sign, resulting in $5 = -9n$. Then, -9 is divided on both sides, leaving $n = -\frac{5}{9}$. This solution can be graphed on the number line with a dot as shown below:

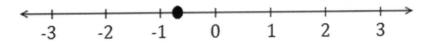

If the problem were changed to say, "The sum of a number and 5 is greater than -8 times the number," then an inequality would be used instead of an equation. Using key words again, *greater than* is represented by the symbol >. The inequality $n + 5 > -8n$ can be solved using the same techniques, resulting in $n < -\frac{5}{9}$. The only time solving an inequality differs from solving an equation is when a negative number is either multiplied by or divided by each side of the inequality. The sign must be switched in this case. For this example, the graph of the solution changes to the following graph because the solution represents all real numbers less than $-\frac{5}{9}$. Not included in this solution is $-\frac{5}{9}$ because it is a *less than* symbol, not *equal to*.

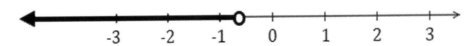

Equations and inequalities in two variables represent a relationship. Jim owns a car wash and charges $40 per car. The rent for the facility is $350 per month. An equation can be written to relate the number of cars Jim cleans to the money he makes per month. Let x represent the number of cars and y represent the profit Jim makes each month from the car wash. The equation $y = 40x - 350$ can be used to show Jim's profit or loss. Since this equation has two variables, the coordinate plane can be used to show the relationship and predict profit or loss for Jim. The following graph shows that Jim must wash at least nine cars to pay the rent, where $x = 9$. Anything nine cars and above yield a profit shown in the value on the y-axis.

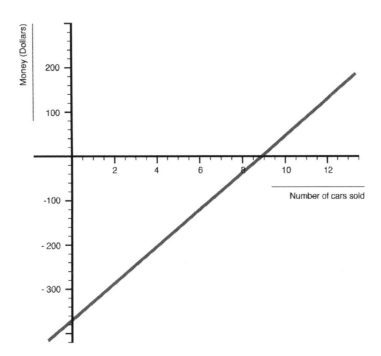

With a single equation in two variables, the solutions are limited only by the situation the equation represents. When two equations or inequalities are used, more constraints are added. For example, in a system of linear equations, there is often—although not always—only one answer. The point of intersection of two lines is the solution. For a system of inequalities, there are infinitely many answers.

The intersection of two solution sets gives the solution set of the system of inequalities. In the following graph, the darker shaded region is where two inequalities overlap. Any set of x and y found in that region satisfies both inequalities. The line with the positive slope is solid, meaning the values on that line are included in the solution.

27

The line with the negative slope is dotted, so the coordinates on that line are not included.

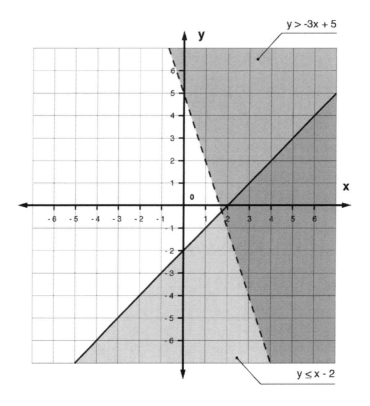

Formulas with two variables are equations used to represent a specific relationship. For example, the formula $d = rt$ represents the relationship between distance, rate, and time. If Bob travels at a rate of 35 miles per hour on his road trip from Westminster to Seneca, the formula $d = 35t$ can be used to represent his distance traveled in a specific length of time. Formulas can also be used to show different roles of the variables, transformed without any given numbers. Solving for r, the formula becomes $\frac{d}{t} = r$. The t is moved over by division so that *rate* is a function of distance and time.

Solving Equations

Solving equations in one variable is the process of isolating that variable on one side of the equation. The letters in an equation and any numbers attached to them are the variables as they stand for unknown quantities that you are trying to solve for. *X* is commonly used as a variable, though any letter can be used. For example, in $3x - 7 = 20$, the variable is $3x$, and it needs to be isolated. The numbers (also called **constants**) are -7 and 20. That means $3x$ needs to be on one side of the equals sign (either side is fine), and all the numbers need to be on the other side of the equals sign.

To accomplish this, the equation must be manipulated by performing opposite operations of what already exists. Remember that addition and subtraction are opposites and that multiplication and division are opposites. Any action taken to one side of the equation must be taken on the other side to maintain equality.

So, since the 7 is being subtracted, it can be moved to the right side of the equation by adding seven to both sides:

$$3x - 7 = 20$$

$$3x - 7 + 7 = 20 + 7$$

$$3x = 27$$

Now that the variable $3x$ is on one side and the constants (now combined into one constant) are on the other side, the 3 needs to be moved to the right side. 3 and x are being multiplied together, so 3 then needs to be divided from each side.

$$\frac{3x}{3} = \frac{27}{3}$$

$$x = 9$$

Now that x has been completely isolated, we know its value.

The solution is found to be $x = 9$. This solution can be checked for accuracy by plugging $x = 9$ in the original equation. After simplifying the equation, $20 = 20$ is found, which is a true statement:

$$3 \times 9 - 7 = 20$$

$$27 - 7 = 20$$

$$20 = 20$$

Equations that require solving for a variable (**algebraic equations**) come in many forms. Here are some more examples:

Sometimes, no number is attached to the variable:

$$x + 8 = 20$$

$$x + 8 - 8 = 20 - 8$$

$$x = 12$$

A fraction may be in the variable:

$$\frac{1}{2}z + 24 = 36$$

$$\frac{1}{2}z + 24 - 24 = 36 - 24$$

$$\frac{1}{2}z = 12$$

Now we multiply the fraction by its inverse:

$$\frac{2}{1} \times \frac{1}{2}z = 12 \times \frac{2}{1}$$

$$z = 24$$

Some algebraic equations contain multiple instances of x:

$$14x + x - 4 = 3x + 2$$

All instances of x can be combined.

$$15x - 4 = 3x + 2$$

$$15x - 4 + 4 = 3x + 2 + 4$$

$$15x = 3x + 6$$

$$15x - 3x = 3x + 6 - 3x$$

$$12x = 6$$

$$\frac{12x}{12} = \frac{6}{12}$$

$$x = \frac{1}{2}$$

Methods for Solving Equations

Equations with one variable can be solved using the addition principle and multiplication principle. If $a = b$, then $a + c = b + c$, and $ac = bc$. Given the equation $2x - 3 = 5x + 7$, the first step is to combine the variable terms and the constant terms. Using the principles, expressions can be added and subtracted onto and off both sides of the equals sign, so the equation turns into $-10 = 3x$. Dividing by 3 on both sides through the multiplication principle with $c = \frac{1}{3}$ results in the final answer of $x = \frac{-10}{3}$.

Some equations have a higher degree and are not solved by simply using opposite operations. When an equation has a degree of 2, completing the square is an option. For example, the quadratic equation $x^2 - 6x + 2 = 0$ can be rewritten by completing the square. The goal of completing the square is to get the equation into the form $(x - p)^2 = q$. Using the example, the constant term 2 first needs to be moved over to the opposite side by subtracting. Then, the square can be completed by adding 9 to both sides, which is the square of half of the coefficient of the middle term $-6x$. The current equation is $x^2 - 6x + 9 = 7$. The left side can be factored into a square of a binomial, resulting in $(x - 3)^2 = 7$. To solve for x, the square root of both sides should be taken, resulting in:

$$(x - 3) = \pm\sqrt{7}$$

$$x = 3 \pm \sqrt{7}$$

Other ways of solving quadratic equations include graphing, factoring, and using the quadratic formula. The equation $y = x^2 - 4x + 3$ can be graphed on the coordinate plane, and the solutions can be observed where it crosses the x-axis. The graph will be a parabola that opens up with two solutions at 1 and 3.

The equation can also be factored to find the solutions. The original equation, $y = x^2 - 4x + 3$ can be factored into $y = (x - 1)(x - 3)$. Setting this equal to zero, the x-values are found to be 1 and 3, just as on the graph. Solving by factoring and graphing are not always possible. The **quadratic formula** is a method of solving quadratic equations that always results in exact solutions.

The formula is:

$$x = \frac{-b \pm \sqrt{b^2 - 4ac}}{2a}$$

where $a, b,$ and c are the coefficients in the original equation in standard form $y = ax^2 + bx + c$. For this example,

$$x = \frac{4 \pm \sqrt{(-4)^2 - 4(1)(3)}}{2(1)} = \frac{4 \pm \sqrt{16 - 12}}{2} = \frac{4 \pm 2}{2} = 1, 3$$

The expression underneath the radical is called the **discriminant**. Without working out the entire formula, the value of the discriminant can reveal the nature of the solutions. If the value of the discriminant $b^2 - 4ac$ is positive, then there will be two real solutions. If the value is zero, there will be one real solution. If the value is negative, the two solutions will be imaginary or complex. If the solutions are complex, it means that the parabola never touches the x-axis. An example of a complex solution can be found by solving the following quadratic: $y = x^2 - 4x + 8$. By using the quadratic formula, the solutions are found to be:

$$x = \frac{4 \pm \sqrt{(-4)^2 - 4(1)(8)}}{2(1)} = \frac{4 \pm \sqrt{16 - 32}}{2} = \frac{4 \pm \sqrt{-16}}{2} = 2 \pm 2i$$

The solutions both have a real part, 2, and an imaginary part, $2i$.

Systems of Equations

A **system of equations** is a group of equations that have the same variables or unknowns. These equations can be linear, but they are not always so. Finding a solution to a system of equations means finding the values of the variables that satisfy each equation. For a linear system of two equations and two variables, there could be a single solution, no solution, or infinitely many solutions.

A single solution occurs when there is one value for x and y that satisfies the system. This would be shown on the graph where the lines cross at exactly one point. When there is no solution, the lines are parallel and do not ever cross. With infinitely many solutions, the equations may look different, but they are the same line. One equation will be a multiple of the other, and on the graph, they lie on top of each other.

The process of elimination can be used to solve a system of equations. For example, the following equations make up a system:

$$x + 3y = 10 \text{ and } 2x - 5y = 9$$

Immediately adding these equations does not eliminate a variable, but it is possible to change the first equation by multiplying the whole equation by -2. This changes the first equation to

$$-2x - 6y = -20$$

The equations can be then added to obtain $-11y = -11$. Solving for y yields $y = 1$. To find the rest of the solution, 1 can be substituted in for y in either original equation to find the value of $x = 7$. The solution to the system is (7, 1) because it makes both equations true, and it is the point in which the lines intersect. If the system is **dependent**—having infinitely many solutions—then both variables will cancel out when the elimination method is used, resulting in an equation that is true for many values of x and y. Since the system is dependent, both equations can be simplified to the same equation or line.

A system can also be solved using **substitution.** This involves solving one equation for a variable and then plugging that solved equation into the other equation in the system. This equation can be solved for one variable, which can then be plugged in to either original equation and solved for the other variable. For example, $x - y = -2$ and $3x + 2y = 9$ can be solved using substitution. The first equation can be solved for x, where $x = -2 + y$. Then it can be plugged into the other equation:

$$3(-2 + y) + 2y = 9$$

Solving for y yields:

$$-6 + 3y + 2y = 9$$

That shows that $y = 3$. If $y = 3$, then $x = 1$.

This solution can be checked by plugging in these values for the variables in each equation to see if it makes a true statement.

Finally, a solution to a system of equations can be found graphically. The solution to a linear system is the point or points where the lines cross. The values of x and y represent the coordinates (x, y) where the lines intersect. Using the same system of equation as above, they can be solved for y to put them in slope-intercept form, $y = mx + b$. These equations become $y = x + 2$ and $y = -\frac{3}{2}x + 4.5$. The slope is the coefficient of x, and the y-intercept is the constant value.

This system with the solution is shown below:

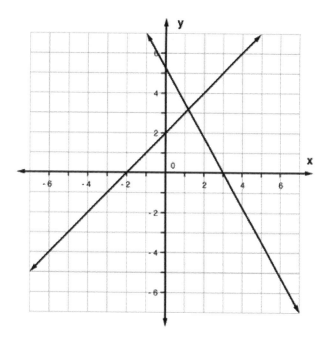

A system of equations may also be made up of a linear and a quadratic equation. These systems may have one solution, two solutions, or no solutions. The graph of these systems involves one straight line and one parabola. Algebraically, these systems can be solved by solving the linear equation for one variable and plugging that answer in to the quadratic equation. If possible, the equation can then be solved to find part of the answer. The graphing method is commonly used for these types of systems. On a graph, these two lines can be found to intersect at one point, at two points across the parabola, or at no points.

Matrices can also be used to solve systems of linear equations. Specifically, for systems, the coefficients of the linear equations in standard form are the entries in the matrix. Using the same system of linear equations as above, $x - y = -2$ and $3x + 2y = 9$, the matrix to represent the system is:

$$\begin{bmatrix} 1 & -1 \\ 3 & 2 \end{bmatrix} \begin{bmatrix} x \\ y \end{bmatrix} = \begin{bmatrix} -2 \\ 9 \end{bmatrix}$$

To solve this system using matrices, the inverse matrix must be found. For a general 2x2 matrix:

$$\begin{bmatrix} a & b \\ c & d \end{bmatrix}$$

The inverse matrix is found by the expression:

$$\frac{1}{ad - bc} \begin{bmatrix} d & -b \\ -c & a \end{bmatrix}$$

The inverse matrix for the system given above is:

$$\frac{1}{2 - -3} \begin{bmatrix} 2 & 1 \\ -3 & 1 \end{bmatrix} = \frac{1}{5} \begin{bmatrix} 2 & 1 \\ -3 & 1 \end{bmatrix}$$

33

The next step in solving is to multiply this identity matrix by the system matrix above. This is given by the following equation:

$$\frac{1}{5}\begin{bmatrix} 2 & 1 \\ -3 & 1 \end{bmatrix}\begin{bmatrix} 1 & -1 \\ 3 & 2 \end{bmatrix}\begin{bmatrix} x \\ y \end{bmatrix} = \begin{bmatrix} -2 \\ 9 \end{bmatrix}\begin{bmatrix} 2 & 1 \\ -3 & 1 \end{bmatrix}\frac{1}{5}$$

which simplifies to

$$\frac{1}{5}\begin{bmatrix} 5 & 0 \\ 0 & 5 \end{bmatrix}\begin{bmatrix} x \\ y \end{bmatrix} = \frac{1}{5}\begin{bmatrix} 5 \\ 15 \end{bmatrix}$$

Solving for the solution matrix, the answer is:

$$\begin{bmatrix} 1 & 0 \\ 0 & 1 \end{bmatrix}\begin{bmatrix} x \\ y \end{bmatrix} = \begin{bmatrix} 1 \\ 3 \end{bmatrix}$$

Since the first matrix is the identity matrix, the solution is $x = 1$ and $y = 3$.

Finding solutions to systems of equations is essentially finding what values of the variables make both equations true. It is finding the input value that yields the same output value in both equations. For functions $g(x)$ and $f(x)$, the equation $g(x) = f(x)$ means the output values are being set equal to each other. Solving for the value of x means finding the x-coordinate that gives the same output in both functions. For example, $f(x) = x + 2$ and $g(x) = -3x + 10$ is a system of equations. Setting $f(x) = g(x)$ yields the equation $x + 2 = -3x + 10$. Solving for x, gives the x-coordinate $x = 2$ where the two lines cross. This value can also be found by using a table or a graph. On a table, both equations can be given the same inputs, and the outputs can be recorded to find the point(s) where the lines cross. Any method of solving finds the same solution, but some methods are more appropriate for some systems of equations than others.

Systems of Linear Inequalities

Systems of linear inequalities are like systems of equations, but the solutions are different. Since inequalities have infinitely many solutions, their systems also have infinitely many solutions. Finding the solutions of inequalities involves graphs. A system of two equations and two inequalities is linear; thus, the lines can be graphed using slope-intercept form. If the inequality has an equals sign, the line is solid. If the inequality only has a greater than or less than symbol, the line on the graph is dotted. Dashed lines indicate that points lying on the line are not included in the solution. After the lines are graphed, a region is shaded on one side of the line. This side is found by determining if a point—known as a **test point**—lying on one side of the line produces a true inequality. If it does, that side of the graph is shaded. If the point produces a false inequality, the line is shaded on the opposite side from the point. The graph of a system of inequalities involves shading the intersection of the two shaded regions.

Properties Involving Algebraic Expressions

Properties such as associativity and commutativity that hold among operations between real numbers also hold between algebraic expressions. Addition and multiplication are associative and commutative; therefore, addition and multiplication can be completed in any order inside an algebraic expression. This is helpful when it comes to solving equations. The addition and multiplication principles state that anything can be added to or multiplied by both sides of an equation to maintain equality. This process is helpful when it comes to isolating the variable. The only time there might be an issue is multiplying by a rational expression with a variable in the denominator. One must make sure that the denominator cannot equal zero. Therefore, it would not be appropriate to multiply both sides of the equation $x^2 = 1$ by $\frac{1}{x}$ to solve for x. The solution $x = 0$ would be lost.

Rate of Change

Rate of change for any line calculates the steepness of the line over a given interval. Rate of change is also known as the slope or rise/run. The rates of change for nonlinear functions vary depending on the interval being used for the function. The rate of change over one interval may be zero, while the next interval may have a positive rate of change. The equation plotted on the graph below, $y = x^2$, is a quadratic function and non-linear. The average rate of change from points $(0, 0)$ to $(1, 1)$ is 1 because the vertical change is 1 over the horizontal change of 1. For the next interval, $(1, 1)$ to $(2, 4)$, the average rate of change is 3 because the slope is $\frac{3}{1}$.

You can see that here:

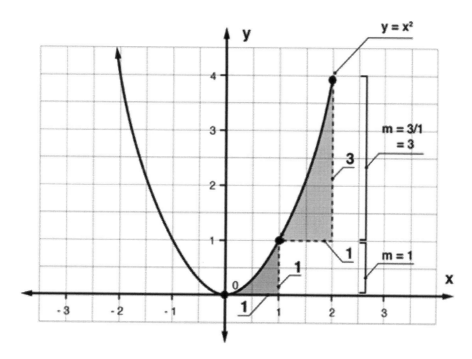

35

The rate of change for a linear function is constant and can be determined based on a few representations. One method is to place the equation in **slope-intercept form**: $y = mx + b$. Thus, m is the slope, and b is the y-intercept. In the graph below, the equation is $y = x + 1$, where the slope is 1 and the y-intercept is 1. For every vertical change of 1 unit, there is a horizontal change of 1 unit. The x-intercept is -1, which is the point where the line crosses the x-axis.

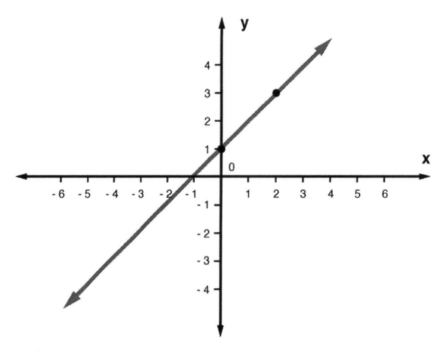

Finding the Zeros of a Function

The **zeros of a function** are the points where its graph crosses the x-axis. At these points, $y = 0$. One way to find the zeros is to analyze the graph. If given the graph, the x-coordinates can be found where the line crosses the x-axis. Another way to find the zeros is to set $y = 0$ in the equation and solve for x. Depending on the type of equation, this could be done by using opposite operations, by factoring the equation, by completing the square, or by using the quadratic formula. If a graph does not cross the x-axis, then the function may have complex roots.

Functions

A **function** is defined as a relationship between inputs and outputs where there is only one output value for a given input. As an example, the following function is in function notation: $f(x) = 3x - 4$. The $f(x)$ represents the output value for an input of x. If $x = 2$, the equation becomes:

$$f(2) = 3(2) - 4 = 6 - 4 = 2$$

The input of 2 yields an output of 2, forming the ordered pair $(2, 2)$. The following set of ordered pairs corresponds to the given function: $(2, 2), (0, -4), (-2, -10)$. The set of all possible inputs of a function is its **domain**, and all possible outputs is called the **range.** By definition, each member of the domain is paired with only one member of the range.

Functions can also be defined **recursively**. In this form, they are not defined **explicitly** in terms of variables. Instead, they are defined using previously-evaluated function outputs, starting with either $f(0)$ or $f(1)$. An example of a recursively-defined function is:

$$f(1) = 2, f(n) = 2f(n - 1) + 2n, n > 1$$

The domain of this function is the set of all integers.

Domain and Range

The domain and range of a function can be found visually by its plot on the coordinate plane. In the function $f(x) = x^2 - 3$, for example, the domain is all real numbers because the parabola can stretch infinitely far left and right with no restrictions. This means that any input value from the real number system will yield an output in the real number system. For the range, the inequality $y \geq -3$ would be used to describe the possible output values because the parabola has a minimum at $y = -3$. This means there will not be any real output values less than -3 because -3 is the lowest value the function reaches on the y-axis.

These same answers for domain and range can be found by observing a table. The table below shows that from input values $x = -1$ to $x = 1$, the output results in a minimum of -3. On each side of $x = 0$, the numbers increase, showing that the range is all real numbers greater than or equal to -3.

x (domain/input)	y (range/output)
-2	1
-1	-2
0	-3
-1	-2
2	1

Function Behavior

Different types of functions behave in different ways. A function is defined to be increasing over a subset of its domain if for all $x_1 \geq x_2$ in that interval, $f(x_1) \geq f(x_2)$. Also, a function is decreasing over an interval if for all $x_1 \geq x_2$ in that interval, $f(x_1) \leq f(x_2)$. A point in which a function changes from increasing to decreasing can also be labeled as the **maximum value** of a function if it is the largest point the graph reaches on the y-axis. A point in which a function changes from decreasing to increasing can be labeled as the **minimum value** of a function if it is the smallest point the graph reaches on the y-axis. Maximum and minimum values are also known as **extreme values**. The graph of a **continuous function** does not have any breaks or jumps in the graph. This description is not true of all functions. A **radical function**, for example, $f(x) = \sqrt{x}$, has a restriction for the domain and range because there are no real negative inputs or outputs for this function. The domain can be stated as $x \geq 0$, and the range is $y \geq 0$.

A **piecewise-defined** function also has a different appearance on the graph. In the following function, there are three equations defined over different intervals. It is a function because there is only one y-value for each x-value, passing the **Vertical Line Test**. The domain is all real numbers less than or equal to 6. The range is all real numbers greater than zero. From left to right, the graph decreases to zero, then increases to almost 4, and then jumps to 6.

From input values greater than 2, the input decreases just below 8 to 4, and then stops.

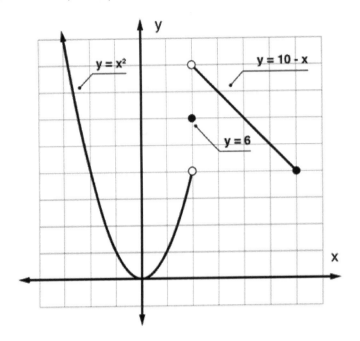

Logarithmic and exponential functions also have different behavior than other functions. These two types of functions are inverses of each other. The **inverse** of a function can be found by switching the place of x and y, and solving for y. When this is done for the exponential equation, $y = 2^x$, the function $y = \log_2 x$ is found. The general form of a **logarithmic function** is $y = \log_b x$, which says b raised to the y power equals x.

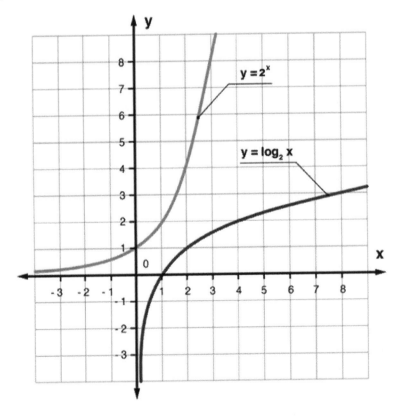

The thick black line on the graph above represents the logarithmic function $y = \log_2 x$. This curve passes through the point $(1, 0)$, just as all log functions do, because any value $b^0 = 1$. The graph of this logarithmic function starts very close to zero, but does not touch the y-axis. The output value will never be zero by the definition of logarithms. The thinner gray line seen above represents the exponential function $y = 2^x$. The behavior of this function is opposite the logarithmic function because the graph of an inverse function is the graph of the original function flipped over the line $y = x$. The curve passes through the point $(0, 1)$ because any number raised to the zero power is one. This curve also gets very close to the x-axis but never touches it because an exponential expression never has an output of zero. The x-axis on this graph is called a horizontal asymptote. An **asymptote** is a line that represents a boundary for a function. It shows a value that the function will get close to, but never reach.

Functions can also be described as being even, odd, or neither. If $f(-x) = f(x)$, the function is **even.** For example, the function $f(x) = x^2 - 2$ is even. Plugging in $x = 2$ yields an output of $y = 2$. After changing the input to $x = -2$, the output is still $y = 2$. The output is the same for opposite inputs. Another way to observe an even function is by the symmetry of the graph. If the graph is symmetrical about the axis, then the function is even. If the graph is symmetric about the origin, then the function is **odd**. Algebraically, if $f(-x) = -f(x)$, the function is odd.

Also, a function can be described as **periodic** if it repeats itself in regular intervals. Common periodic functions are trigonometric functions. For example, $y = \sin x$ is a periodic function with period 2π because it repeats itself every 2π units along the x-axis.

Building a Function

Functions can be built out of the context of a situation. For example, the relationship between the money paid for a gym membership and the months that someone has been a member can be described through a function. If the one-time membership fee is \$40 and the monthly fee is \$30, then the function can be written $f(x) = 30x + 40$. The x-value represents the number of months the person has been part of the gym, while the output is the total money paid for the membership. The table below shows this relationship. It is a representation of the function because the initial cost is \$40 and the cost increases each month by \$30.

x (months)	y (money paid to gym)
0	40
1	70
2	100
3	130

Functions can also be built from existing functions. For example, a given function $f(x)$ can be transformed by adding a constant, multiplying by a constant, or changing the input value by a constant. The new function $g(x) = f(x) + k$ represents a vertical shift of the original function. In $f(x) = 3x - 2$, a vertical shift 4 units up would be:

$$g(x) = 3x - 2 + 4 = 3x + 2$$

Multiplying the function by a constant k represents a vertical stretch, based on whether the constant is greater than or less than 1. The function

$$g(x) = kf(x) = 4(3x - 2) = 12x - 8$$

represents a stretch. Changing the input x by a constant forms the function:

$$g(x) = f(x + k) = 3(x + 4) - 2 = 3x + 12 - 2 = 3x + 10$$

and this represents a horizontal shift to the left 4 units. If $(x - 4)$ was plugged into the function, it would represent a vertical shift.

A **composition function** can also be formed by plugging one function into another. In function notation, this is written:

$$(f \circ g)(x) = f(g(x))$$

For two functions $f(x) = x^2$ and $g(x) = x - 3$, the composition function becomes:

$$f(g(x)) = (x - 3)^2 = x^2 - 6x + 9$$

The composition of functions can also be used to verify if two functions are inverses of each other. Given the two functions $f(x) = 2x + 5$ and $g(x) = \frac{x-5}{2}$, the composition function can be found $(f \circ g)(x)$. Solving this equation yields:

$$f(g(x)) = 2\left(\frac{x - 5}{2}\right) + 5 = x - 5 + 5 = x$$

It also is true that $g(f(x)) = x$. Since the composition of these two functions gives a simplified answer of x, this verifies that $f(x)$ and $g(x)$ are inverse functions. The domain of $f(g(x))$ is the set of all x-values in the domain of $g(x)$ such that $g(x)$ is in the domain of $f(x)$. Basically, both $f(g(x))$ and $g(x)$ have to be defined.

To build an inverse of a function, $f(x)$ needs to be replaced with y, and the x and y values need to be switched. Then, the equation can be solved for y. For example, given the equation $y = e^{2x}$, the inverse can be found by rewriting the equation $x = e^{2y}$. The natural logarithm of both sides is taken down, and the exponent is brought down to form the equation:

$$\ln(x) = \ln(e)\, 2y$$

$\ln(e)$=1, which yields the equation $\ln(x) = 2y$. Dividing both sides by 2 yields the inverse equation

$$\frac{\ln(x)}{2} = y = f^{-1}(x)$$

The domain of an inverse function is the range of the original function, and the range of an inverse function is the domain of the original function. Therefore, an ordered pair (x, y) on either a graph or a table corresponding to $f(x)$ means that the ordered pair (y, x) exists on the graph of $f^{-1}(x)$. Basically, if $f(x) = y$, then $f^{-1}(y) = x$. For a function to have an inverse, it must be one-to-one. That means it must pass the **Horizontal Line Test,** and if any horizontal line passes through the graph of the function twice, a function is not one-to-one. The domain of a function that is not one-to-one can be restricted to an interval in which the function is one-to-one, to be able to define an inverse function.

Functions can also be formed from combinations of existing functions.

Given $f(x)$ and $g(x)$, the following can be built:

$$f + g$$

$$f - g$$

$$fg$$

$$\frac{f}{g}$$

The domains of $f + g$, $f - g$, and fg are the intersection of the domains of f and g. The domain of $\frac{f}{g}$ is the same set, excluding those values that make $g(x) = 0$.

For example, if:

$$f(x) = 2x + 3$$

$$g(x) = x + 1$$

then

$$\frac{f}{g} = \frac{2x + 3}{x + 1}$$

Its domain is all real numbers except -1.

Common Functions

Three common functions used to model different relationships between quantities are linear, quadratic, and exponential functions. **Linear functions** are the simplest of the three, and the independent variable x has an exponent of 1. Written in the most common form, $y = mx + b$, the coefficient of x tells how fast the function grows at a constant rate, and the b-value tells the starting point. A **quadratic function** has an exponent of 2 on the independent variable x. Standard form for this type of function is $y = ax^2 + bx + c$, and the graph is a parabola. These type functions grow at a changing rate. **An exponential function** has an independent variable in the exponent $y = ab^x$. The graph of these types of functions is described as **growth** or **decay**, based on whether the **base**, b, is greater than or less than 1. These functions are different from quadratic functions because the base stays constant. A common base is base e.

The following three functions model a linear, quadratic, and exponential function respectively: $y = 2x$, $y = x^2$, and $y = 2^x$. Their graphs are shown below. The first graph, modeling the linear function, shows that the growth is constant over each interval. With a horizontal change of 1, the vertical change is 2. It models a constant positive growth. The second graph shows the quadratic function, which is a curve that is symmetric across the y-axis. The growth is not constant, but the change is mirrored over the axis. The last graph models the exponential function, where the horizontal change of 1 yields a vertical change that increases more and more. The exponential graph gets very close to the x-axis, but never touches it, meaning there is an asymptote there. The y-value can never be zero because the base of 2 can never be raised to an input value that yields an output of zero.

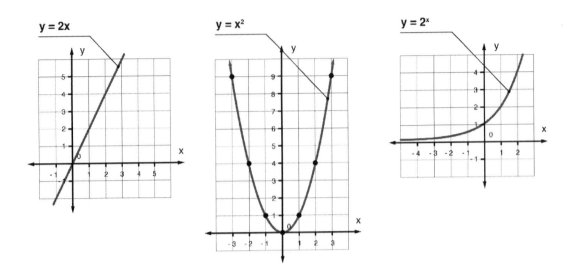

The three tables below show specific values for three types of functions. The third column in each table shows the change in the y-values for each interval. The first table shows a constant change of 2 for each equal interval, which matches the slope in the equation $y = 2x$. The second table shows an increasing change, but it also has a pattern. The increase is changing by 2 more each time, so the change is quadratic. The third table shows the change as factors of the base, 2. It shows a continuing pattern of factors of the base.

y = 2x		
x	y	Δy
1	2	
2	4	2
3	6	2
4	8	2
5	10	2

y = x²		
x	y	Δy
1	1	
2	4	3
3	9	5
4	16	7
5	25	9

y = 2ˣ		
x	y	Δy
1	2	
2	4	2
3	8	4
4	16	8
5	32	16

Given a table of values, the type of function can be determined by observing the change in y over equal intervals. For example, the tables below model two functions. The changes in interval for the x-values is 1 for both tables. For the first table, the y-values increase by 5 for each interval. Since the change is constant, the situation can be described as a linear function. The equation would be $y = 5x + 3$. For the second table, the change for y is 5, 20, 100, and 500, respectively. The increases are multiples of 5, meaning the situation can be modeled by an exponential function. The equation $y = 5^x + 3$ models this situation.

x	y
0	3
1	8
2	13
3	18
4	23

x	y
0	3
1	8
2	28
3	128
4	628

Quadratic equations can be used to model real-world area problems. For example, a farmer may have a rectangular field that he needs to sow with seed. The field has length $x + 8$ and width $2x$. The formula for area should be used: $A = lw$. Therefore:

$$A = (x + 8) \times 2x = 2x^2 + 16x$$

The possible values for the length and width can be shown in a table, with input x and output A. If the equation was graphed, the possible area values can be seen on the y-axis for given x-values.

Exponential growth and decay can be found in real-world situations. For example, if a piece of notebook paper is folded 25 times, the thickness of the paper can be found. To model this situation, a table can be used. The initial point is one-fold, which yields a thickness of 2 papers. For the second fold, the thickness is 4. Since the thickness doubles each time, the table below shows the thickness for the next few folds. Notice the thickness changes by the same factor each time. Since this change for a constant interval of folds is a factor of 2, the function is exponential. The equation for this is $y = 2^x$. For twenty-five folds, the thickness would be 33,554,432 papers.

x (folds)	y (paper thickness)
0	1
1	2
2	4
3	8
4	16
5	32

One exponential formula that is commonly used is the **interest formula**: $A = Pe^{rt}$. In this formula, interest is compounded continuously. A is the value of the investment after the time, t, in years. P is the initial amount of the investment, r is the interest rate, and e is the constant equal to approximately 2.718. Given an initial amount of $200 and a time of 3 years, if interest is compounded continuously at a rate of 6%, the total investment value can be found by plugging each value into the formula. The invested value at the end is $239.44. In more complex problems, the final investment may be given, and the rate may be the unknown. In this case, the formula becomes $239.44 = 200e^{r3}$. Solving for r requires isolating the exponential expression on one side by dividing by 200, yielding the equation

$1.20 = e^{r3}$. Taking the natural log of both sides results in $\ln(1.2) = r3$. Using a calculator to evaluate the logarithmic expression, $r = 0.06 = 6\%$.

When working with logarithms and exponential expressions, it is important to remember the relationship between the two. In general, the logarithmic form is $y = log_b x$ for an exponential form $b^y = x$. Logarithms and exponential functions are inverses of each other.

Trigonometric Functions

Trigonometric functions are also used to describe behavior in mathematics. **Trigonometry** is the relationship between the angles and sides of a triangle. **Trigonometric functions** include sine, cosine, tangent, secant, cosecant, and cotangent. The functions are defined through ratios in a right triangle. **SOHCAHTOA** is a common acronym used to remember these ratios, which are defined by the relationships of the sides and angles relative to the right angle. **Sine** is opposite over hypotenuse, **cosine** is adjacent over hypotenuse, and **tangent** is opposite over adjacent. These ratios are the reciprocals of secant, cosecant, and cotangent, respectively. Angles can be measured in degrees or radians. Here is a diagram of SOHCAHTOA:

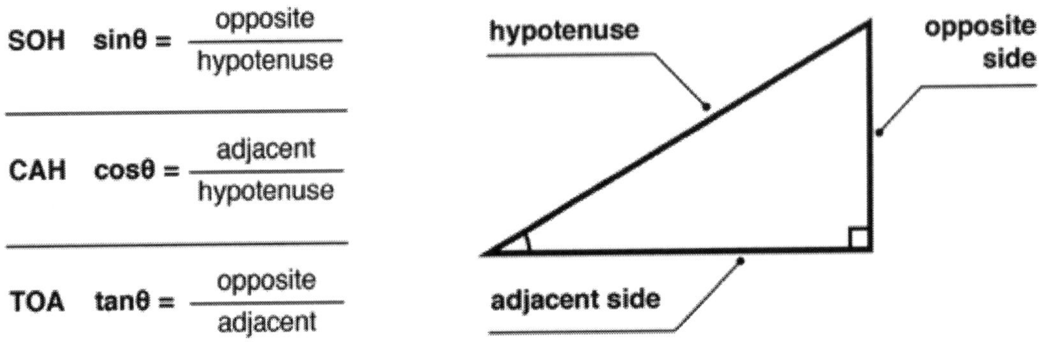

A **radian** is equal to the angle that subtends the arc with the same length as the radius of the circle. It is another unit for measuring angles, in addition to degrees. The **unit circle** is used to describe different radian measurements and the trigonometric ratios for special angles. The circle has a center at the origin, $(0, 0)$, and a radius of 1, which can be seen below. The points where the circle crosses an axis are labeled.

The circle begins on the right-hand side of the x-axis at 0 radians. Since the circumference of a circle is $2\pi r$ and the radius $r = 1$, the circumference is 2π. Zero and 2π are labeled as radian measurements at the point $(1, 0)$ on the graph. The radian measures around the rest of the circle are labeled also in relation to π; π is at the point $(-1, 0)$, also known as 180 degrees. Since these two measurements are equal, $\pi = 180$ degrees written as a ratio can be used to convert degrees to radians or vice versa. For example, to convert 30 degrees to radians, 30 degrees $\times \frac{\pi}{180 \text{ degrees}}$ can be used to obtain $\frac{1}{6}\pi$ or $\frac{\pi}{6}$. This radian measure is a point the unit circle

The coordinates labeled on the unit circle are found based on two common right triangles. The ratios formed in the coordinates can be found using these triangles. Each of these triangles can be inserted into the circle to correspond 30, 45, and 60 degrees or $\frac{\pi}{6}, \frac{\pi}{4}$, and $\frac{\pi}{3}$ radians.

By letting the hypotenuse length of these triangles equal 1, these triangles can be placed inside the unit circle. These coordinates can be used to find the trigonometric ratio for any of the radian measurements on the circle.

Given any (x, y) on the unit circle, $\sin(\theta) = y$, $\cos(\theta) = x$, and $\tan(\theta) = \frac{y}{x}$. The value θ is the angle that spans the arc around the unit circle. For example, finding $\sin(\frac{\pi}{4})$ means finding the y-value corresponding to the angle $\theta = \frac{\pi}{4}$. The answer is $\frac{\sqrt{2}}{2}$. Finding $\cos(\frac{\pi}{3})$ means finding the x-value corresponding to the angle $\theta = \frac{\pi}{3}$. The answer is $\frac{1}{2}$ or 0.5. Both angles lie in the first quadrant of the unit circle. Trigonometric ratios can also be calculated for radian measures past $\frac{\pi}{2}$, or 90 degrees. Since the same special angles can be moved around the circle, the results only differ with a change in sign. This can be seen at two points labeled in the second and third quadrant.

Trigonometric functions are periodic. Both sine and cosine have period 2π. For each input angle value, the output value follows around the unit circle. Once it reaches the starting point, it continues around and around the circle. It is true that:

$$\sin(0) = \sin(2\pi) = \sin(4\pi), \text{etc.}$$

and

$$\cos(0) = \cos(2\pi) = \cos(4\pi)$$

Tangent has period π, and its output values repeat themselves every half of the unit circle. The domain of sine and cosine are all real numbers, and the domain of tangent is all real numbers, except the points where cosine equals zero. It is also true that

$$\sin(-x) = -\sin x$$

$$\cos(-x) = \cos(x)$$

$$\tan(-x) = -\tan(x)$$

Thus, sine and tangent are odd functions, while cosine is an even function. Sine and tangent are symmetric with respect the origin, and cosine is symmetric with respect to the y-axis.

The graph of trigonometric functions can be used to model different situations. General forms are

$$y = a \sin b(x - h) + k$$

and

$$y = a \cos b (x - h) + k$$

The variable a represents the **amplitude**, which shows the maximum and minimum value of the function. The b is used to find the **period** by using the ratio $\frac{2\pi}{b}$, h is the **horizontal shift**, and k is the **vertical shift**.

The equation $y = \sin(x)$ is shown on the following graph with the thick black line. The stretched graph of $y = 2\sin(x)$ is shown in solid black, and the shrunken graph $y = \frac{1}{2}\sin(x)$ is shown with the dotted line.

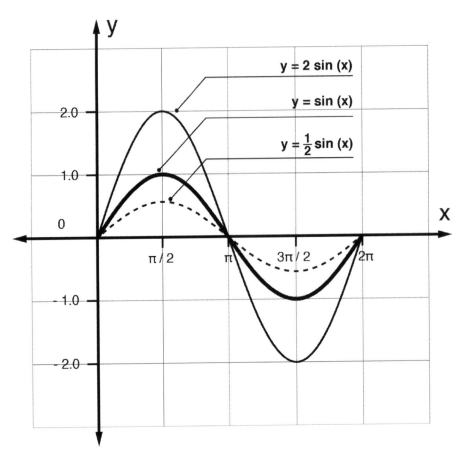

Trigonometric functions are used to find unknown ratios for a given angle measure. The inverse of these trig functions is used to find the unknown angle, given a ratio. For example, the expression $\arcsin(\frac{1}{2})$ means finding the value of x for $\sin(x) = \frac{1}{2}$. Since $\sin(\theta) = \frac{y}{1}$ on the unit circle, the angle whose y-value is $\frac{1}{2}$ is $\frac{\pi}{6}$. The inverse of any of the trigonometric functions can be used to find a missing angle measurement. Values not found on the unit circle can be found using the trigonometric functions on the calculator, making sure its mode is set to degrees or radians.

In order for the inverse to exist, the function must be one-to-one over its domain. There cannot be two input values connected to the same output. For example, the following graphs show the functions $y = \cos(x)$ and $y = \arccos(x)$. In order to have an inverse, the domain of cosine is restricted from 0 to π.

Therefore, the range of its inverse function is $[0, \pi]$.

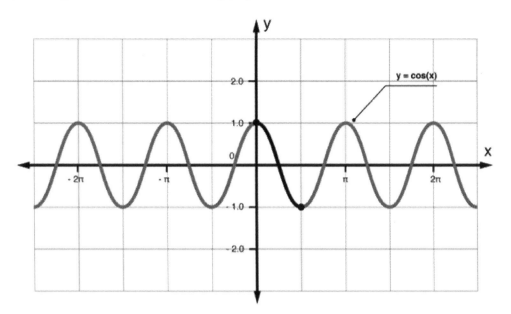

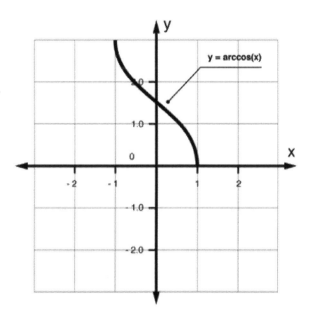

Inverses of trigonometric functions can be used to solve real-world problems. For example, there are many situations where the lengths of a perceived triangle can be found, but the angles are unknown. Consider a problem where the height of a flag (25 feet) and the distance on the ground to the flag is given (42 feet). The unknown, x, is the angle. To find this angle, the equation $\tan x = \frac{42}{25}$ is used. To solve for x, the inverse function can be used to turn the equation into $\tan^{-1} \frac{42}{25} = x$. Using the calculator, in degree mode, the answer is found to be $x = 59.2$ degrees

Trigonometric Identities

From the unit circle, the trigonometric ratios were found for the special right triangle with a hypotenuse of 1.

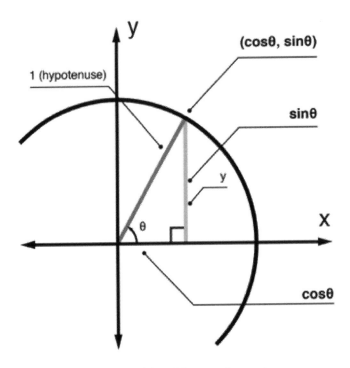

From this triangle, the following **Pythagorean identities** are formed:

$$\sin^2 \theta + \cos^2 \theta = 1$$

$$\tan^2 \theta + 1 = \sec^2 \theta$$

$$1 + \cot^2 \theta = \csc^2 \theta$$

The second two identities are formed by manipulating the first identity. Since identities are statements that are true for any value of the variable, then they may be used to manipulate equations. For example, a problem may ask for simplification of the expression $\cos^2 x + \cos^2 x \tan^2 x$. Using the fact that

$$\tan(x) = \frac{\sin x}{\cos x}$$

$\frac{\sin^2 x}{\cos^2 x}$ can then be substituted in for $\tan^2 x$, making the expression:

$$\cos^2 x + \cos^2 x \frac{\sin^2 x}{\cos^2 x}$$

Then the two $\cos^2 x$ terms on top and bottom cancel each other out, simplifying the expression to:

$$\cos^2 x + \sin^2 x$$

By the first Pythagorean identity stated above, the expression can be turned into:

$$\cos^2 x + \sin^2 x = 1$$

Another set of trigonometric identities are the **double-angle formulas**:

$$\sin 2\alpha = 2 \sin \alpha \, \cos \alpha$$

$$\cos 2\alpha = \begin{cases} \cos^2\alpha - \sin^2\alpha \\ 2\cos^2\alpha - 1 \\ 1 - 2\sin^2\alpha \end{cases}$$

Using these formulas, the following identity can be proved:

$$\sin 2x = \frac{2 \tan x}{1 + \tan^2 x}$$

By using one of the Pythagorean identities, the denominator can be rewritten as:

$$1 + \tan^2 x = \sec^2 x$$

By knowing the reciprocals of the trigonometric identities, the secant term can be rewritten to form the equation:

$$\sin 2x = \frac{2 \tan x}{1} * \cos^2 x$$

Replacing $\tan(x)$, the equation becomes:

$$\sin 2x = \frac{2 \sin x}{\cos x} * \cos^2 x$$

The $\cos x$ can cancel out. The new equation is:

$$\sin 2x = 2 \sin x * \cos x$$

This final equation is one of the double-angle formulas.

Other trigonometric identities such as half-angle formulas, sum and difference formulas, and difference of angles formulas can be used to prove and rewrite trigonometric equations. Depending on the given equation or expression, the correct identities need to be chosen to write equivalent statements.

The graph of sine is equal to the graph of cosine, shifted $\frac{\pi}{2}$ units. Therefore, the function $y = \sin x$ is equal to:

$$y = \cos(\frac{\pi}{2} - x)$$

Within functions, adding a constant to the independent variable shifts the graph either left or right. By shifting the cosine graph, the curve lies on top of the sine function. By transforming the function, the two equations give the same output for any given input.

Functions of Two Variables

The graph of a function of one variable can be represented in the xy-plane and is known as a **curve**. When a function has two variables, the function is graphed in three-dimensional space, and the graph is known as a **surface.** The graph is the set of all ordered triples (x, y, z) that satisfy the function. Within three-dimensional space, there is a third axis known as the **z-axis.**

Solving Trigonometric Functions

Solving trigonometric functions can be done with a knowledge of the unit circle and the trigonometric identities. It requires the use of opposite operations combined with trigonometric ratios for special triangles. For example, the problem may require solving the equation $2\cos^2 x - \sqrt{3}\cos x = 0$ for the values of x between 0 and 180 degrees. The first step is to factor out the $\cos x$ term, resulting in:

$$\cos x \left(2\cos x - \sqrt{3}\right) = 0$$

By the factoring method of solving, each factor can be set equal to zero:

$$\cos x = 0$$

$$(2\cos x - \sqrt{3}) = 0$$

The second equation can be solved to yield the following equation:

$$\cos x = \frac{\sqrt{3}}{2}$$

Now that the value of x is found, the trigonometric ratios can be used to find the solutions of $x = 30$ and 90 degrees.

Solving trigonometric functions requires the use of algebra to isolate the variable and a knowledge of trigonometric ratios to find the value of the variable. The unit circle can be used to find answers for special triangles. Beyond those triangles, a calculator can be used to solve for variables within the trigonometric functions.

Solving Logarithmic and Exponential Functions
To solve an equation involving exponential expressions, the goal is to isolate the exponential expression. Once this process is completed, the logarithm—with the base equaling the base of the exponent of both sides—needs to be taken to get an expression for the variable. If the base is e, the natural log of both sides needs to be taken.

To solve an equation with logarithms, the given equation needs to be written in exponential form, using the fact that $\log_b y = x$ means $b^x = y$, and then solved for the given variable. Lastly, properties of logarithms can be used to simplify more than one logarithmic expression into one.

Calculus

Limits of Functions

The **limit** of a function can be described as the output that is approached as the input approaches a certain value. Written in function notation, the limit of $f(x)$ as x approaches a is $\lim_{x \to a} f(x) = B$. As x draws near to some value a, represented by $x \to a$, then $f(x)$ approaches some number B. In the graph of the function $f(x) = \frac{x+2}{x+2}$, the line is continuous except where $x = -2$. Because $x = -2$ yields an undefined output and a hole in the graph, the function does not exist at this value. The limit, however, does exist. As the value $x = -2$ is approached from the left side, the output is getting very close to 1. From the right side, as the x-value approaches -2, the output gets close to 1 also. Since the function value from both sides approaches 1, then:

$$\lim_{x \to -2} \frac{x+2}{x+2} = 1$$

One special type of function, the **step function** $f(x) = [x]$, can be used to define right and left-hand limits. The graph is shown below. The left-hand limit as x approaches 1 is $\lim_{x \to 1^-}[x]$. From the graph, as x approaches 1 from the left side, the function approaches 0. For the right-hand limit, the expression is $\lim_{x \to 1^+}[x]$. The value for this limit is one. Since the function does not have the same limit for the left and right side, then the limit does not exist at $x = 1$. From that same reasoning, the limit does not exist for any integer for this function.

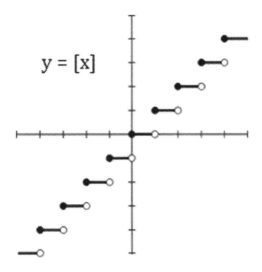

$$y = [x]$$

Sometimes a function approaches infinity as it draws near to a certain x-value. For example, the following graph shows the function $f(x) = \frac{2x}{x-3}$. There is an asymptote at $x = 3$. The limit as x approaches 3, $\lim_{x \to 3} \frac{2x}{x-3}$, does not exist. The right and left-hand side limits at 3 do not approach the same output value. One approaches positive infinity, and the other approaches negative infinity. **Infinite limits** do not satisfy the definition of a limit.

The limit of the function as x approaches a number must be equal to a finite value.

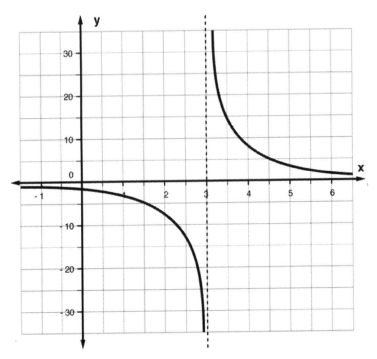

Horizontal asymptotes can be found using limits. Horizontal asymptotes are limits as x approaches either ∞ or $-\infty$. For example, to find $\lim\limits_{x \to \infty} \frac{2x}{x-3}$, the graph can be used to see the value of the function as x grows larger and larger. For this example, the limit is 2, so it has a horizontal asymptote of $y = 2$. In considering $\lim\limits_{x \to -\infty} \frac{2x}{x-3} = 2$, the limits can also be seen on a graphing calculator by plotting the equation $y = \frac{2x}{x-3}$. Then the table can be brought up. By scrolling up and down, the limit can be found as x approaches any value.

Limit laws exist that assist in finding limits of functions. These properties include multiplying by a constant:

$$\lim k f(x) = k \lim f(x)$$

And the addition property:

$$\lim[f(x) + g(x)] = \lim f(x) + \lim g(x)$$

Two other properties are the multiplication property:

$$\lim f(x)g(x) = (\lim f(x))(\lim g(x))$$

And the division property:

$$\lim \frac{f(x)}{g(x)} = \frac{\lim f(x)}{\lim g(x)} \ (if \lim g(x) \neq 0)$$

These properties are helpful in finding limits of polynomial functions algebraically.

In the following equation, the constant and multiplication properties can be used together, and the problem can be rewritten:

$$\lim_{x \to 2} 4x^2 - 3x + 8$$

$$4\lim_{x \to 2} x^2 - \lim_{x \to 2} 3x + \lim_{x \to 2} 8$$

Since this is a continuous function, direct substitution can be used. The value of 2 is substituted in for x and evaluated as $4(2^2) - 3(2) + 8$, which yields a limit of 18. These properties allow functions to be rewritten so that limits can be calculated.

Derivatives

The **derivative** of a function is found using the limit of the difference quotient:

$$\lim_{\Delta x \to 0} \frac{f(x + \Delta x) - f(x)}{\Delta x}$$

This finds the slope of the tangent line of the given function at a given point. It is the slope, $\frac{\Delta y}{\Delta x}$, as $\Delta x \to$ 0. The derivative can be denoted in many ways, such as $f'(x)$, y', or $\frac{dy}{dx}$.

The following graph plots a function in black. The gray line represents a **secant line**, formed between two chosen points on the graph. The slope of this line can be found using rise over run. As these two points get closer to zero, meaning Δx approaches 0, the tangent line is found. The slope of the tangent line is equal to the limit of the slopes of the secant lines as $\Delta x \to 0$.

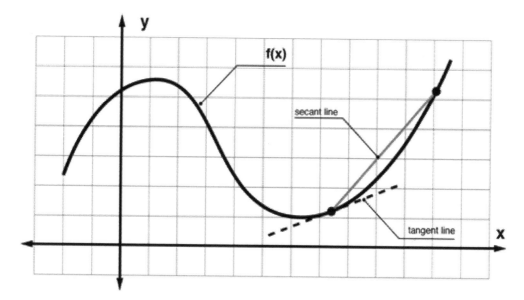

The derivative of a function can be found algebraically using the limit definition. Here is the process for finding the derivative of $f(x) = x^2 - 2$:

$$f'(x) = \lim_{h \to 0} \frac{f(x+h) - f(x)}{h}$$

$$= \lim_{h \to 0} \frac{(x+h)^2 - 2 - (x^2 - 2)}{h}$$

$$= \lim_{h \to 0} \frac{(x+h)(x+h) - 2 - x^2 + 2}{h}$$

$$= \lim_{h \to 0} \frac{x^2 + xh + xh + h^2 - 2 - x^2 + 2}{h}$$

$$= \lim_{h \to 0} \frac{x^2 + 2xh + h^2 - 2 - x^2 + 2}{h}$$

$$= \lim_{h \to 0} \frac{2xh + h^2}{h}$$

$$= \lim_{h \to 0} \frac{h(2x + h)}{h} = \lim_{h \to 0} 2x + h = 2x + 0 = 2x$$

Once the derivative function is found, it can be evaluated at any point by substituting that value in for x. Therefore, in this example, $f'(2) = 4$.

Continuity

To find if a function is continuous, the definition consists of three steps. These three steps include finding $f(a)$, finding $\lim_{x \to a} f(x)$, and finding $\lim_{x \to a} f(x) = f(a)$. If the limit of a function equals the function value at that point, then the function is continuous at $x = a$. For example, the function $f(x) = \frac{1}{x}$ is continuous everywhere except $x = 0$. $f(0) = \frac{1}{0}$ is undefined; therefore, the function is discontinuous at 0. Secondly, to determine if the function $f(x) = \frac{1}{x-1}$ is continuous at 2, its function value must equal its limit at 2. First,

$$f(2) = \frac{1}{2 - 1} = 1$$

Then the limit can be found by direct substitution:

$$\lim_{x \to 2} \frac{1}{x - 1} = 1$$

Since these two values are equal, then the function is continuous at:

$$x = 2$$

Differentiability and continuity are related in that if the derivative can be found at $x = c$, then the function is continuous at $x = c$. If the slope of the tangent line can be found at a certain point, then there is no hole or jump in the graph at that point. Some functions, however, can be continuous while not differentiable at a given point. An example is the graph of the function $f(x) = |x|$. At the origin, the

54

derivative does not exist, but the function is still continuous. Points where a function is discontinuous are where a vertical tangent exists and where there is a cusp or corner at a given x-value.

Estimating Derivatives and Integrals

Since the derivative is a slope, a table of values can be used to approximate the derivative. The change in y divided by the change in x gives the slope at a point. Using the points in the table, slopes of secant lines can be calculated. Based on the limit of those slopes, the derivative at that point can be approximated. Take the following table for example:

x	$f(x)$
0	0
1	1
4	2
9	3
16	4

To find $f'(4)$ using the table, the slope between $f(4)$ and each of the other points needs to be calculated. The following table shows the slopes between different points. Based on the slopes found from the table, the value of $f'(4)$ is between 0.2 and 0.5.

Given Point	Point to find Secant Line	Slope
(4, 2)	(0, 0)	0.5
(4, 2)	(1, 1)	0.333333
(4, 2)	(9, 3)	0.2
(4, 2)	(16, 4)	0.16667

An **integral** is the antiderivative, and the integral of $f(x)$ is denoted as $\int f(x)dx$. An integral can be explained through the following equation:

$$\frac{dy}{dx} \int_a^x f(t)dt = f(x)$$

Taking the derivative of the integral of a function yields the original function. On a graph, integrals find the area under a curve.

One way to estimate integrals is by the **trapezoid rule**. Over a defined integral, the area under the curve can be split up into trapezoids. These shapes come close to covering the area under the curve. Once split into trapezoids, the area of each shape is found, and then all areas are added together. The following graph offers an example of how to use the trapezoid rule. The defined integral of the function from 1 to 5 is split into four trapezoids. For each trapezoid, the width is 1. The two lengths can then be averaged. These two numbers are multiplied together to find the area.

Once all four areas are calculated, the approximate integral can be found:

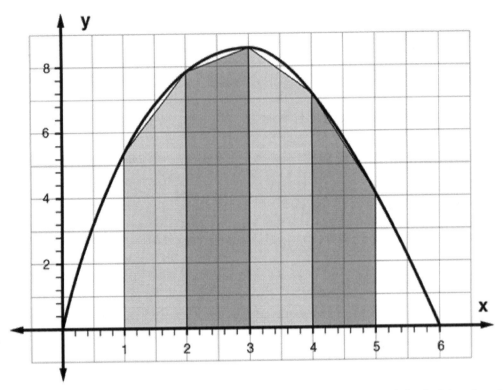

The **midpoint rule** uses rectangles instead of trapezoids to calculate area, and the height of each rectangle equals the function evaluated at the midpoint of each subinterval.

Differentiation and Integration Techniques

Finding the derivative of a function can be done using the definition as described above, but rules proved via the different quotient can also be used. A few are listed below. These rules apply for functions that take the form inside the parenthesis. For example, the function $f(x) = 3x^4$ would use the **Power Rule** and **Constant Multiple Rule**. To find the derivative, the exponent is brought down to be multiplied by the coefficient, and the new exponent is one less than the original. As an equation, the derivative is $f'(x) = 12x^3$.

$$\frac{d}{dx}(a^b) = 0$$

$$\frac{d}{dx}(x^n) = nx^{n-1}$$

$$\frac{d}{dx}(a^x) = a^x \ln a$$

$$\frac{d}{dx}(x^x) = x^x(1 + \ln x)$$

In relation to real-life problems, the position of a ball that is thrown into the area may be given by the equation $p = 7 + 25t - 16t^2$. The position, p, can be found for any time, t, after the ball is thrown. To find the initial position, $t = 0$ can be substituted into the equation to find p. That position would be 7ft above the ground, which is equal to the constant at the end of the equation.

Finding the derivative of the function would use the Power Rule. The derivative is $p' = 25 - 32t$. The derivative of a position function represents the velocity function. To find the initial velocity, the time $t = 0$ can be substituted into the equation. The initial velocity is found to be 25ft/s – the same as the coefficient of t in the position equation. Taking the derivative of the velocity equation yields the acceleration equation $p'' = -32$. This value is the acceleration at which a ball is pulled by gravity to the ground in feet per second squared.

Since integration is the inverse operation of finding the derivative, the integral is found by going backwards from the derivative. In relation to the ball problem, an acceleration function can be integrated to find the velocity function. That function can then be integrated to find the position function. From velocity, integration finds the position function $p = -16t^2 + 25t + c$, where c is an unknown constant. More information would need to be given in the original problem to integrate and find the value of c.

Behavior of a Function

Derivatives can be used to find the behavior of different functions such as the extrema, concavity, and symmetry. Given a function $f(x) = 3x^2$, the **first derivative** is $f'(x) = 6x$. This equation describes the slope of the line. Setting the derivative equal to zero means finding where the slope is zero, and these are potential points in which the function has extreme values. If the first derivative is positive over an interval, the function is increasing over that interval. If the first derivative is negative over an interval, the function is decreasing over that interval. Therefore, if the derivative is equal to zero at a point and the function changes from increasing to decreasing, then the function has a minimum at that point. If the function changes from decreasing to increasing at that point, it is a maximum. The **second derivative** can be used to define concavity. If it is positive over an interval, the graph resembles a U and is concave up over that interval. If the second derivative is negative, the graph is concave down. For this equation, solving $f'(x) = 6x = 0$ gets $x = 0$, $f(0) = 0$. Also, the second derivative is 6, which is positive. The graph is concave up and, therefore, has a minimum value at (0,0).

Foundational Theorems of Calculus

Per the **fundamental theorem of calculus**, on a closed interval [a, b], the following represents the definite integral: $f(x)$:

$$\int_a^b f(x)dx = F(b) - F(a)$$

$F(x)$ represents the antiderivative of the function $f(x)$. Other theorems allow constants to be moved to the front of the integral, negatives to be moved to the outside of the integral, and integrals to be split into two parts that make up a whole. An example of using these theorems can be seen in the following problem:

$$\int_{-1}^3 (4x^3 - 2x)dx = (108 - 6) - (-4 + 2) = 104$$

The antiderivative of $4x^3 - 2x$ is $x^4 - x^2$.

Within the fundamental theorem of calculus, the **antiderivative** $F(x)$ exists. It is true that $F'(x) = f(x)$. Therefore, it is important to know how the graph of a function and a derivative relate. Because the derivative function represents the slope of the tangent of a function—where a function is horizontal—the derivative function has zeros. On intervals where the function is decreasing, the derivative function lies below the x-axis, and on intervals where the function is increasing, the derivative function lies above the y-axis.

Slope is defined in algebra to be a rate of change; therefore, the derivative function is a rate of change. The definite integral in the fundamental theorem of calculus can also be used to represent a rate of change. If one were to calculate the definite integral of a function $f(x)$ over the interval $[a, b]$ as $F(b) - F(a)$, where $F'(x) = f(x)$, the result is the net rate of change of $F(x)$ over the same interval.

The **average value of a function** can be found by the following integral:

$$\frac{1}{b-a}\int_a^b f(x)dx$$

The integral finds the area of the region bounded by the function and the x-axis, while the fraction divides the area to find the average value of the integral. An example of this is shown in the graph below. The function $f(x)$ is the black line. The light gray shading represents the area under the curve, while the rectangle drawn on top with added darker shading represents the same amount of area as the region under the graph of the given function over the interval of a to b. This rectangle has the base [a, b] and height f(c).

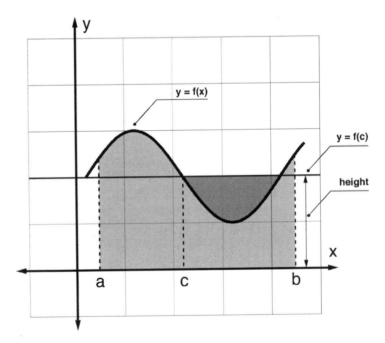

The **Mean Value Theorem** states that if f is a continuous function on interval [a,b], and f' is differentiable on (a,b), then there exists at least one number, c, in which the derivative at that point equals the slope of the secant line connecting the endpoints of the interval.

This number can be found by the equation:

$$f'(c) = \frac{f(b) - f(a)}{b - a}$$

Riemann Sums

Riemann sums can be used to calculate the area under a curve $y = f(x)$ from $x = a$ to $x = b$. In other words, Riemann sums can be used to find $\int_a^b f(x)dx$. The interval from a to b is divided into n subintervals of equal length Δx, and the function is evaluated at a point x_i^* in each interval. This creates a rectangle over each subinterval, and the area of each rectangle can be found and summed. The area of each rectangle is:

$$f(x_i^*)\Delta x$$

The total area is:

$$\sum_{i=1}^{n} f(x_i^*)\Delta x$$

If there are infinitely many subintervals, the limit of this expression can be taken as $n \to \infty$ to represent the definite integral.

Differentiation and Integration in Real-World Problems

Integration can be described as an accumulation process because it takes many small areas and adds them up to find a total area over an interval. That process can be used in many real-world problems that deal with volume, area, and distance. Differentiation can also be used in real-world problems. For example, a company may want to maximize the size of the boxes it uses to ship its products. The boxes are to be cut out of a piece of cardboard that measures 8 inches long and 5 inches tall. Since squares must be cut out of the corners to make the boxes, the size of the square needs to be altered to maximize the box volume. The volume of a box can be found using the formula $V = l \times w \times h$. For length, the expression is $(8 - 2x)$ because the initial length is 8, and length x is taken from both sides of the original length to form the box. The width expression is $(5 - 2x)$. The height of the box is x. Therefore, the volume function is

$$V = (8 - 2x)(5 - 2x)x = 40x - 26x^2 + 4x^3$$

Taking the derivative, $V' = 40 - 52x + 12x^2$, and setting it equal to zero will find the potential maximum and minimum points. If a maximum is found, the x-value represents the amount that needs to be cut from the corners of the box to maximize the volume. To find the volume at its max, the x-value can be substituted into the original equation.

Sequences and Series

A **sequence** is an enumerated set of numbers, and each term or member is defined by the number that exists within the sequence. It can have either a finite or infinite number of terms, and a sequence is written as $\{a_n\}$, where a_n is the nth term of the sequence.

An example of an infinite sequence is:

$$\left\{\frac{n+1}{n^2}\right\}_{n=1}^{\infty}$$

Its first three terms are found by evaluating at n=1, 2, and 3 to get 2, $\frac{3}{4}$, and $\frac{4}{9}$. Limits of infinite sequences, if they exist, can be found in a similar manner as finding infinite limits of functions. n needs to be treated as a variable, and then $\lim\limits_{n\to\infty} \frac{n+1}{n^2}$ can be evaluated, resulting in 0.

An **infinite series** is the sum of an infinite sequence. For example, $\sum_{n=1}^{\infty} \frac{n+1}{n^2}$ is the infinite series of the sequence given above. Partial sums are sums of a finite number of terms. For example, s_{10} represents the sum of the first 10 terms, and in general, s_n represents the sum of the first n terms. An infinite series can either converge or diverge. If the sum of an infinite series is a finite number, the series is said to **converge**. Otherwise, it **diverges**. In the general infinite series $\sum a_n$, If $\lim\limits_{n\to\infty} a_n \neq 0$ or does not exist, the series diverges. However, if $\lim\limits_{n\to\infty} a_n = 0$, the series does not necessarily converge.

Several tests exist that determine whether a series converges:

- The **Absolute Convergence Test** states that if $\sum |a_n|$ converges, then $\sum a_n$ converges.

- The **Integral Test** states that if $f(n) = a_n$ is a positive, continuous, decreasing function, then $\sum a_n$ is convergent if and only if $\int_1^{\infty} f(x)dx$ is convergent. The geometric series $\sum ar^{n-1}$ is **convergent** if $|r| < 1$ and its sum is equal to $\frac{a}{1-r}$. If $|r| \geq 1$, the geometric series is **divergent**.

- The **Limit Comparison Tests** compares two infinite series $\sum a_n$ and $\sum b_n$ with positive terms. If $\sum b_n$ converges and $a_n \leq b_n$ for all n, then $\sum a_n$ converges. If $\sum b_n$ diverges and $a_n \geq b_n$ for all n, then $\sum a_n$ diverges. If $\lim\limits_{n\to\infty} \frac{a_n}{b_n} = c$, where c is a finite, positive number, then either both series converge or diverge.

- The **Alternating Series Test** states that if $b_{n+1} \geq b_n$ for all n and $\lim\limits_{n\to\infty} b_n = 0$, then the series $\sum (-1)^{n-1} b_n$.

- The **Ratio Test** states that if the limit of the ratio of consecutive terms a_{n+1}/a_n is less than 1, then the series is convergent. If the ratio is greater than 1, the series is divergent. If the limit is equal to 1, the test is inconclusive.

- The **Root Test** states that if $\lim\limits_{n\to\infty} \sqrt[n]{|a_n|} < 1$, the series converges. If the same limit is greater than 1, the series diverges, and if the limit equals 1, the test is inconclusive.

Practice Questions

1. Which of the following is the result of simplifying the expression:
$$\frac{4a^{-1}b^3}{a^4b^{-2}} \times \frac{3a}{b}$$

    a. $12a^3b^5$

    b. $12\frac{b^4}{a^4}$

    c. $\frac{12}{a^4}$

    d. $7\frac{b^4}{a}$

2. If x is not zero, then $\frac{3}{x} + \frac{5u}{2x} - \frac{u}{4} =$

    a. $\frac{12+10u-ux}{4x}$

    b. $\frac{3+5u-ux}{x}$

    c. $\frac{12x+10u+ux}{4x}$

    d. $\frac{12+10u-u}{4x}$

3. The function $f(x) = (x-2)^3$ satisfies the hypotheses of the Mean Value Theorem on the interval $[0,2]$. What number c that satisfies the theorem?

    a. $2 + \frac{2\sqrt{3}}{3}$

    b. $-2 - \frac{2\sqrt{3}}{3}$

    c. 0

    d. 2

4. Which of the following is a factor of both $x^2 + 4x + 4$ and $x^2 - x - 6$?

    a. $x - 3$

    b. $x + 2$

    c. $x - 2$

    d. $x + 3$

5. Which of the following augmented matrices represents the system of equations below?

$$2x - 3y + z = -5$$
$$4x - y - 2z = -7$$
$$-x + 2z = -1$$

a. $\begin{bmatrix} 2 & -3 & 1 & -5 \\ 4 & -1 & -2 & -7 \\ -1 & 0 & 2 & -1 \end{bmatrix}$

b. $\begin{bmatrix} 2 & 4 & -1 \\ -3 & -1 & 0 \\ 1 & -2 & 2 \\ -5 & -7 & -1 \end{bmatrix}$

c. $\begin{bmatrix} 2 & 4 & -1 & -5 \\ -3 & -1 & 0 & -7 \\ 2 & -2 & 2 & -1 \end{bmatrix}$

d. $\begin{bmatrix} 2 & -3 & 1 \\ 4 & -1 & -2 \\ -1 & 0 & 2 \end{bmatrix}$

6. What are the zeros of the function: $f(x) = x^3 + 4x^2 + 4x$?
   a. -2
   b. 0, -2
   c. 2
   d. 0, 2

7. If $g(x) = x^3 - 3x^2 - 2x + 6$ and $f(x) = 2$, then what is $g(f(x))$?
   a. -26
   b. 6
   c. $2x^3 - 6x^2 - 4x + 12$
   d. -2

8. What is the solution to the following system of equations?

$$x^2 - 2x + y = 8$$
$$x - y = -2$$

   a. $(-2, 3)$
   b. There is no solution.
   c. $(-2, 0)\ (1, 3)$
   d. $(-2, 0)\ (3, 5)$

9. Which of the following shows the correct result of simplifying the following expression:

$$(7n + 3n^3 + 3) + (8n + 5n^3 + 2n^4)$$

   a. $9n^4 + 15n - 2$
   b. $2n^4 + 5n^3 + 15n - 2$
   c. $9n^4 + 8n^3 + 15n$
   d. $2n^4 + 8n^3 + 15n + 3$

10. What is the product of the following expression?
$$(4x - 8)(5x^2 + x + 6)$$
    a. $20x^3 - 36x^2 + 16x - 48$
    b. $6x^3 - 41x^2 + 12x + 15$
    c. $204 + 11x^2 - 37x - 12$
    d. $2x^3 - 11x^2 - 32x + 20$

11. How could the following equation be factored to find the zeros?
$$y = x^3 - 3x^2 - 4x$$
    a. $0 = x^2(x - 4), x = 0, 4$
    b. $0 = 3x(x + 1)(x + 4), x = 0, -1, -4$
    c. $0 = x(x + 1)(x + 6), x = 0, -1, -6$
    d. $0 = x(x + 1)(x - 4), x = 0, -1, 4$

12. What is the simplified quotient of the following expression?
$$\frac{5x^3}{3x^2y} \div \frac{25}{3y^9}$$
    a. $\dfrac{125x}{9y^{10}}$

    b. $\dfrac{x}{5y^8}$

    c. $\dfrac{5}{xy^8}$

    d. $\dfrac{xy^8}{5}$

13. What is the solution for the following equation?
$$\frac{x^2 + x - 30}{x - 5} = 11$$
    a. $x = -6$
    b. There is no solution.
    c. $x = 16$
    d. $x = 5$

14. Mom's car drove 72 miles in 90 minutes. How fast did she drive in feet per second?
    a. 0.8 feet per second
    b. 48.9 feet per second
    c. 0.009 feet per second
    d. 70.4 feet per second

15. How do you solve $V = lwh$ for h?
    a. $lwV = h$
    b. $h = \dfrac{V}{lw}$
    c. $h = \dfrac{Vl}{w}$
    d. $h = \dfrac{Vw}{l}$

16. What is the domain for the function $y = \sqrt{x}$?

    a. All real numbers

    b. $x \geq 0$

    c. $x > 0$

    d. $y \geq 0$

17. What is the answer to $(2 + 2i)(2 - 2i)$?

    a. 8

    b. $8i$

    c. 4

    d. $4i$

18. What is the most general antiderivative of the function: $g(x) = \frac{1+x+x^2}{\sqrt{x}}$?

    a. $G(x) = x^{\frac{1}{2}} + x^{\frac{3}{2}} + x^{\frac{5}{2}} + c$

    b. $G(x) = 2x^{\frac{1}{2}} + \frac{2}{3}x^{\frac{3}{2}} + \frac{2}{5}x^{\frac{5}{2}} + c$

    c. $G(x) = 2x^{\frac{1}{2}} + \frac{2}{3}x^{\frac{3}{2}} + \frac{2}{5}x^{\frac{5}{2}}$

    d. $G(x) = x^{\frac{1}{2}} + x^{\frac{3}{2}} + x^{\frac{5}{2}}$

19. What is the function that forms an equivalent graph to $y = \cos(x)$?

    a. $y = \tan(x)$

    b. $y = \csc(x)$

    c. $y = \sin(x + \frac{\pi}{2})$

    d. $y = \sin(x - \frac{\pi}{2})$

20. What is the solution for the equation $\tan(x) + 1 = 0$, where $0 \leq x < 2\pi$?

    a. $x = \frac{3\pi}{4}, \frac{5\pi}{4}$

    b. $x = \frac{3\pi}{4}, \frac{\pi}{4}$

    c. $x = \frac{5\pi}{4}, \frac{7\pi}{4}$

    d. $x = \frac{3\pi}{4}, \frac{7\pi}{4}$

21. What is the inverse of the function $f(x) = 3x - 5$?

    a. $f^{-1}(x) = \frac{x}{3} + 5$

    b. $f^{-1}(x) = \frac{5x}{3}$

    c. $f^{-1}(x) = 3x + 5$

    d. $f^{-1}(x) = \frac{x+5}{3}$

22. What are the zeros of $f(x) = x^2 + 4$?

    a. $x = -4$

    b. $x = \pm 2i$

    c. $x = \pm 2$

    d. $x = \pm 4i$

23. For which of the following are $x = 4$ and $x = -4$ solutions?
    a. $x^2 + 16 = 0$
    b. $x^2 + 4x - 4 = 0$
    c. $x^2 - 2x - 2 = 0$
    d. $x^2 - 16 = 0$

24. What is the simplified form of the expression $1.2 \times 10^{12} \div 3.0 \times 10^8$?
    a. 0.4×10^4
    b. 4.0×10^4
    c. 4.0×10^3
    d. 3.6×10^{20}

25. You measure the width of your door to be 36 inches. The true width of the door is 35.75 inches. What is the relative error in your measurement?
    a. 0.7%
    b. 0.007%
    c. 0.99%
    d. 0.1%

26. What are the y-intercept(s) for $y = x^2 + 3x - 4$?
    a. $y = 1$
    b. $y = -4$
    c. $y = 3$
    d. $y = 4$

27. Is the following function even, odd, neither, or both?
$$y = \frac{1}{2}x^4 + 2x^2 - 6$$
    a. Even
    b. Odd
    c. Neither
    d. Both

28. Which equation is not a function?
    a. $y = |x|$
    b. $y = \sqrt{x}$
    c. $x = 3$
    d. $y = 4$

29. How could the following function be rewritten to identify the zeros?
$$y = 3x^3 + 3x^2 - 18x$$
    a. $y = 3x(x + 3)(x - 2)$
    b. $y = x(x - 2)(x + 3)$
    c. $y = 3x(x - 3)(x + 2)$
    d. $y = (x + 3)(x - 2)$

30. What is the slope of the line tangent to the graph of $y = x^3 - 4$ at the point where $x = 2$?

    a. $3x^2$

    b. 4

    c. -4

    d. 12

31. Let $f(x) = \begin{cases} \frac{x^2-4}{x-2} & if\ x \neq 2 \\ 0 & if\ x = 2 \end{cases}$. Which of the following statements is/are true?

    I. $\lim_{x \to 2} exists$

    II. $f(2) exists$

    III. $f\ is\ continuous\ at\ x = 2$

    a. I only

    b. II only

    c. I and II

    d. I and III

32. What is $\lim_{x \to 4} \frac{x^2-16}{x-4}$?

    a. 0

    b. 1

    c. 8

    d. Nonexistent

33. What are the first four terms of the following series?
$$\left\{ \frac{(-1)^{n+1}}{n^2 + 5} \right\}_{n=0}^{\infty}$$

    a. $\frac{1}{6}, \frac{1}{9}, \frac{1}{14}, \frac{1}{19}$

    b. $\frac{1}{6}, \frac{-1}{9}, \frac{1}{14}, \frac{-1}{19}$

    c. $\frac{-1}{5}, \frac{1}{6}, \frac{-1}{9}, \frac{1}{14}$

    d. $\frac{1}{5}, \frac{1}{6}, \frac{1}{9}, \frac{1}{14}$

34. A particle moves along the x-axis so that at any time $t \geq 0$, its velocity is given by $v(t) = \frac{6}{t+3}$. What is the acceleration of the particle at time $t = 5$?

    a. $-\frac{2}{3}$

    b. $-\frac{3}{32}$

    c. $\frac{3}{4}$

    d. $\frac{2}{3}$

35. Given the graph of the derivative of $f(x)$, on what interval(s) is the graph of $f(x)$ increasing?

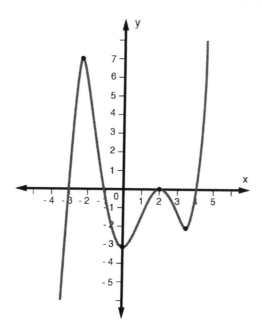

a. $(-3, -1)(4, \infty)$
b. $(-\infty, -2.4)(0, 2)(3.4, \infty)$
c. $(-\infty, -3)(-1, 4)$
d. $(0, \infty)$

36. What is the trapezoidal approximation for the integral $\int_0^4 \sqrt{x} \, dx$, using 4 subintervals?
   a. 5.333
   b. 12.293
   c. 10.293
   d. 5.146

37. What is the definite integral that represents the area of the region bounded by the graphs of $y_1 = 5 - x^2$ and $y_2 = -3x - 5$?

a. $\int_{-\sqrt{5}}^{\sqrt{5}} (5 - x^2) \, dx$

b. $\int_{-\sqrt{5}}^{\sqrt{5}} (x^2 - 3x - 10) \, dx$

c. $\int_{-2}^{5} (-x^2 + 3x + 10) \, dx$

d. $\int_{-2}^{5} [(5 - x^2) + (-3x - 5)] \, dx$

38. What type of function is modeled by the values in the following table?

X	f(x)
1	2
2	4
3	8
4	16
5	32

   a. Linear
   b. Exponential
   c. Quadratic
   d. Cubic

39. What is the simplified form of the expression $tan\theta \; cos\theta$?
   a. $sin\theta$
   b. 1
   c. $csc\theta$
   d. $\dfrac{1}{sec\theta}$

40. An investment of $2,000 is made into an account with an annual interest rate of 5%, compounded continuously. What is the total value for the investment after eight years?
   a. $4,707
   b. $3,000
   c. $2,983.65
   d. $10, 919.63

41. Is the series $\sum_{k=0}^{\infty}(-1)^k \left(\frac{2}{3}\right)^k$ convergent or divergent? If convergent, find its sum.
   a. Divergent
   b. Convergent, $\dfrac{3}{5}$
   c. Convergent, $\dfrac{5}{3}$
   d. Convergent, $\dfrac{2}{3}$

42. What is the sum of $(2.6 \times 10^5) + (1.3 \times 10^4)$? Enter your answer in the answer box below.

43. What would the equation be for the following problem? Enter your answer in the answer box below.
   3 times the sum of a number and 7 is greater than or equal to 32.

44. What are the zeros of the function: $x^3 + 2x^2 - 15x$? Select all that apply.
   a. 0
   b. -3
   c. -5
   d. 3

68

Answer Explanations

1. B: To simplify the given equation, the first step is to make all exponents positive by moving them to the opposite place in the fraction. This expression becomes $\frac{4b^3 b^2}{a^1 a^4} \times \frac{3a}{b}$. Then the rules for exponents can be used to simplify. Multiplying the same bases means the exponents can be added. Dividing the same bases means the exponents are subtracted.

2. A: The common denominator here will be $4x$. Rewrite these fractions as $\frac{3}{x} + \frac{5u}{2x} - \frac{u}{4} = \frac{12}{4x} + \frac{10u}{4x} - \frac{ux}{4x} = \frac{12x+10u-ux}{4x}$.

3. A: The Mean Value Theorem states that because the function is continuous and differentiable on the given interval, there exists a number, c, in the given interval (0,2) that satisfies $f'(c) = \frac{f(2)-f(0)}{2-0}$. Therefore, $3(c-2)^2 = 4$. Placing this quadratic equation in standard form gives $3x^2 - 12x + 8 = 0$. The quadratic formula yields $c = 2 \pm \frac{2\sqrt{3}}{3}$. Only the root within the given interval of (0, 2) satisfies the theorem.

4. B: To factor $x^2 + 4x + 4$, the numbers needed are those that add to 4 and multiply to 4. Therefore, both numbers must be 2, and the expression factors to $x^2 + 4x + 4 = (x+2)^2$. Similarly, the expression factors to $x^2 - x - 6 = (x-3)(x+2)$, so that they have $x+2$ in common.

5. A: The augmented matrix that represents the system of equations has dimensions 4×3 because there are three equations with three unknowns. The coefficients of the variables make up the first three columns, and the last column is made up of the numbers to the right of the equals sign. This system can be solved by reducing the matrix to row-echelon form, where the last column gives the solution for the unknown variables.

6. B: There are two zeros for the given function. They are $x = 0, -2$. The zeros can be found a number of ways, but this particular equation can be factored into $f(x) = x(x^2 + 4x + 4) = x(x+2)(x+2)$. By setting each factor equal to zero and solving for x, there are two solutions. On a graph, these zeros can be seen where the line crosses the x-axis.

7. D: This problem involves a composition function, where one function is plugged into the other function. In this case, the $f(x)$ function is plugged into the $g(x)$ function for each x-value. The composition equation becomes $g(f(x)) = 2^3 - 3(2^2) - 2(2) + 6$. Simplifying the equation gives the answer $g(f(x)) = 8 - 3(4) - 2(2) + 6 = 8 - 12 - 4 + 6 = -2$.

8. D: This system of equations involves one quadratic function and one linear function, as seen from the degree of each equation. One way to solve this is through substitution. Solving for y in the second equation yields $y = x + 2$. Plugging this equation in for the y of the quadratic equation yields $x^2 - 2x + x + 2 = 8$. Simplifying the equation, it becomes $x^2 - x + 2 = 8$. Setting this equal to zero and factoring, it becomes $x^2 - x - 6 = 0 = (x-3)(x+2)$. Solving these two factors for x gives the zeros $x = 3, -2$. To find the y-value for the point, each number can be plugged in to either original equation. Solving each one for y yields the points $(3, 5)$ and $(-2, 0)$.

9. D: The expression is simplified by collecting like terms. Terms with the same variable and exponent are like terms, and their coefficients can be added.

10. A: Finding the product means distributing one polynomial to the other so that each term in the first is multiplied by each term in the second. Then, like terms can be collected. Multiplying the factors yields the expression $20x^3 + 4x^2 + 24x - 40x^2 - 8x - 48$. Collecting like terms means adding the x^2 terms and adding the x terms. The final answer after simplifying the expression is $20x^3 - 36x^2 + 16x - 48$.

11. D: Finding the zeros for a function by factoring is done by setting the equation equal to zero, then completely factoring. Since there was a common x for each term in the provided equation, that is factored out first. Then the quadratic that is left can be factored into two binomials: $(x + 1)(x - 4)$. Setting each factor equation equal to zero and solving for x yields three zeros.

12. D: Dividing rational expressions follows the same rule as dividing fractions. The division is changed to multiplication, and the reciprocal is found in the second fraction. This turns the expression into $\frac{5x^3}{3x^2} * \frac{3y^9}{25}$. Multiplying across and simplifying, the final expression is $\frac{xy^8}{5}$.

13. B: The equation can be solved by factoring the numerator into $(x + 6)(x - 5)$. Since that same factor $(x - 5)$ exists on top and bottom, that factor cancels. This leaves the equation $x + 6 = 11$. Solving the equation gives the answer $x = 5$. When this value is plugged into the equation, it yields a zero in the denominator of the fraction. Since this is undefined, there is no solution.

14. D: This problem can be solved by using unit conversions. The initial units are miles per minute. The final units need to be feet per second. Converting miles to feet uses the equivalence statement 1 mile=5,280 feet. Converting minutes to seconds uses the equivalence statement 1 minute=60 seconds. Setting up the ratios to convert the units is shown in the following equation: $\frac{72 \, miles}{90 \, minutes} * \frac{1 \, minute}{60 \, seconds} * \frac{5280 \, feet}{1 \, mile} = 70.4$ feet per second. The initial units cancel out, and the new, desired units are left.

15. B: The formula can be manipulated by dividing both sides by the length, l, and the width, w. The length and width will cancel on the right, leaving height by itself.

16. B: The domain is all possible input values, or x-values. For this equation, the domain is every number greater than or equal to zero. There are no negative numbers in the domain because taking the square root of a negative number results in an imaginary number.

17. A: This answer is correct because $(2 + 2i)(2 - 2i)$, using the FOIL method and rules for imaginary numbers, is:

$$4 - 4i + 4i - 4i^2 = 8$$

Choice *B* is not the answer because there is no *i* in the final answer, since the *i*'s cancel out in the FOIL. Choice *C*, 4, is not the final answer because we add $4 + 4$ in the end to equal 8. Choice *D*, 4*i*, is not the final answer because there is neither a 4 nor an *i* in the final answer.

18. B: First, the function can be rewritten as $g(x) = x^{-\frac{1}{2}} + x^{\frac{1}{2}} + x^{\frac{3}{2}}$. The antiderivative is found by using the rule that $\int x^n = \frac{x^{n+1}}{n+1} + c$. Therefore, $G(x) = 2x^{\frac{1}{2}} + \frac{2}{3}x^{\frac{3}{2}} + \frac{2}{5}x^{\frac{5}{2}} + c$. Only one constant (+c) is necessary for it to be the most general antiderivative.

19. C: Graphing the function $y = \cos(x)$ shows that the curve starts at $(0, 1)$, has an amplitude of 2, and a period of 2π. This same curve can be constructed using the sine graph, by shifting the graph to the left $\frac{\pi}{2}$ units. This equation is in the form $y = \sin(x + \frac{\pi}{2})$.

20. D: Using SOHCAHTOA, tangent is $\frac{y}{x}$ for the special triangles. Since the value needs to be negative one, the angle must be some form of 45 degrees or $\frac{\pi}{4}$. The value is negative in the second and fourth quadrant, so the answer is $\frac{3\pi}{4}$ and $\frac{7\pi}{4}$.

21. D: The inverse of a function is found by following these steps:

1. Change f(x) to y.

2. Switch the x and y in the equation.

3. Solve for y. In the given equation, solving for y is done by adding 5 to both sides, then dividing both sides by 3.

This answer can be checked on the graph by verifying the lines are reflected over $y = x$.

22. B: The zeros of this function can be found by using the quadratic formula:

$$x = \frac{-b \pm \sqrt{b^2 - 4ac}}{2a}$$

Identifying *a*, *b*, and *c* can be done from the equation as well because it is in standard form. The formula becomes:

$$x = \frac{0 \pm \sqrt{0^2 - 4(1)(4)}}{2(1)} = \frac{\sqrt{-16}}{2}$$

Since there is a negative underneath the radical, the answer is a complex number.

23. D: There are two ways to approach this problem. Each value can be substituted into each equation. A can be eliminated, since $4^2 + 16 = 32$. Choice B can be eliminated, since $4^2 + 4 \cdot 4 - 4 = 28$. C can be eliminated, since $4^2 - 2 \cdot 4 - 2 = 6$. But, plugging in either value into $x^2 - 16$, which gives $(\pm 4)^2 - 16 = 16 - 16 = 0$.

24. C: Scientific notation division can be solved by grouping the first terms together and grouping the tens together. The first terms can be divided, and the tens terms can be simplified using the rules for exponents. The initial expression becomes $0.4 * 10^4$. This is not in scientific notation because the first number is not between 1 and 10. Shifting the decimal and subtracting one from the exponent, the answer becomes $4.0 * 10^3$.

25. A: The relative error can be found by finding the absolute error and making it a percent of the true value. The absolute value is $36 - 35.75 = 0.25$. This error is then divided by 36—the true value—to find 0.7%.

26. B: The y-intercept of an equation is found where the x-value is zero. Plugging zero into the equation for x, the first two terms cancel out, leaving -4.

27. A: The equation is *even* because $f(-x) = f(x)$. Plugging in a negative value will result in the same answer as when plugging in the positive of that same value. The function:

$$f(-2) = \frac{1}{2}(-2)^4 + 2(-2)^2 - 6 = 8 + 8 - 6 = 10$$

yields the same value as:

$$f(2) = \frac{1}{2}(2)^4 + 2(2)^2 - 6 = 8 + 8 - 6 = 10$$

28. C: The equation $x = 3$ is not a function because it does not pass the vertical line test. This test is made from the definition of a function, where each x-value must be mapped to one and only one y-value. This equation is a vertical line, so the x-value of 3 is mapped with an infinite number of y-values.

29. A: The function can be factored to identify the zeros. First, the term $3x$ is factored out to the front because each term contains $3x$. Then, the quadratic is factored into $(x + 3)(x - 2)$.

30. D: Finding the slope of the line tangent to the given function involves taking the derivative twice. The first derivative gives the line tangent to the graph. The second derivative finds the slope of that line. The line tangent to the graph has an equation $y' = 3x^2$. The slope of this line at $x = 2$ is found by the second derivative, $y = 6x$, or $y = 6(2) = 12$.

31. C: The limit exists because $\lim_{x \to 2} f(x) = 4$. The limit as x approaches two is four, and the function value $f(2) = 0$; thus, they are not equal. Because they are not the same, the function is not continuous, and the first and second statements are the only ones that are true.

32. C: The numerator can be factored into $(x + 4)(x - 4)$. Since there is a factor of $(x - 4)$ in the numerator and denominator, these factors cancel, leaving the $(x + 4)$. Plugging in $x = 4$ into this function yields $4 + 4 = 8$.

33. C: The numerator in this sequence indicates that the sign of each term changes from term to term:

$$\left\{ \frac{(-1)^{n+1}}{n^2 + 5} \right\}_{n=0}^{\infty}$$

The first term is negative because $n = 0$ and $-1^{n+1} = -1^1 = -1$. Therefore, the second term is positive. The third term is negative, etc. The denominator is evaluated like a function for plugging in the various n value. For example, the denominator of the first term, when n = 0, is $0^2 + 5 = 0$.

34. B: The acceleration of the particle can be found by taking the derivative of the velocity equation. This equation is:

$$v'(t) = \frac{0 - 6(1)}{(t + 3)^2} = \frac{-6}{(t + 3)^2}$$

Finding the acceleration at time $t = 5$ can be found by plugging five in for the variable t in the derivative. The equation and answer are:

$$v'(5) = \frac{-6}{(5 + 3)^2} = \frac{-6}{64} = \frac{-3}{32}$$

35. A: The graph of $f'(x)$ is positive on the intervals $(-3, -1)$ and $(4, \infty)$.

36. D: The graph of $f(x) = \sqrt{x}$ can be split into four trapezoids to approximate the integral. These subintervals can be represented in this expression:

$$\frac{4 - 0}{2(4)}[f(0) + 2f(1) + 2f(2) + 2f(3) + f(4)]$$

The area of a trapezoid is equal to one-half the sum of the two bases multiplied by the height. Each function value in the brackets represents the sum of the bases of the trapezoids. The fraction outside the brackets represents the height of the trapezoids, dividing by two for the one-half. Finding the function values and simplifying the expression leads to the answer 5.146.

37. C: Setting the y-values of each equation equal to one another finds the point where they meet. The equation $5 - x^2 = -3x - 5$ can be simplified by solving for 0, $0 = x^2 - 3x - 10$. This equation can be factored into $0 = (x + 2)(x - 5)$. The zeros are $x = -2$ and $x = 5$, between $x = -2$ and $x = 5$, $y_1 > y_2$.

38. B: The table shows values that are increasing exponentially. The differences between the inputs are the same, while the differences in the outputs are changing by a factor of 2. The values in the table can be modeled by the equation $f(x) = 2^x$.

39. A: Using the trigonometric identity $\tan(\theta) = \frac{\sin(\theta)}{\cos(\theta)}$, the expression becomes $\frac{\sin\theta}{\cos\theta}\cos\theta$. The factors that are the same on the top and bottom cancel out, leaving the simplified expression $\sin\theta$.

40. C: The formula for continually compounded interest is $A = Pe^{rt}$. Plugging in the given values to find the total amount in the account yields the equation $A = 2000e^{0.05*8} = 2983.65$.

41. B: The given series is a geometric series because it can be written as $\sum_{k=0}^{\infty} \left(\frac{-2}{3}\right)^k$, and it is convergent because $|r| = \frac{2}{3} < 1$. Its sum is:

$$\frac{1}{1 - (-\frac{2}{3})} = \frac{3}{5}$$

42. 2.73×10^5: The exponent of the ten must be the same before any operations are performed on the numbers. So, $(2.6 * 10^5) + (1.3 * 10^4)$ cannot be added until one of the exponents on the ten is changed. The $1.3 * 10^4$ can be changed to $0.13 * 10^5$, then the 2.6 and 0.13 can be added. The answer comes out to be $2.73 * 10^5$.

43. $3(n + 7) \geq 32$: 3 times the sum of a number and 7 is greater than or equal to 32 can be translated into equation form utilizing mathematical operators and numbers.

44. A, C, and D: There are three zeros for the given function. They are $x = 0, 3, -5$. The zeros can be found a number of ways, but this particular equation can be factored into $f(x) = x(x^2 + 2x - 15) = x(x - 3)(x + 5)$. By setting each factor equal to zero and solving for x, there are three solutions. On a graph, these zeros can be seen where the line crosses the x-axis.

Geometry, Probability & Statistics, and Discrete Mathematics

Geometry

Transformations in a Plane

Angles, Circles, Line Segments, Perpendicular Lines, and Parallel Lines

In geometry, a **line** connects two points, has no thickness, and extends indefinitely in both directions beyond each point. If the length is finite, it's known as a **line segment** and has two **endpoints.** A **ray** is the straight portion of a line that has one endpoint and extends indefinitely in the other direction. An **angle** is formed when two rays begin at the same endpoint and extend indefinitely. The endpoint of an angle is called a **vertex. Adjacent angles** are two side-by-side angles formed from the same ray that have the same endpoint. Angles are measured in **degrees** or **radians,** which is a measure of **rotation**. A full rotation equals 360 degrees or 2π radians, which represents a circle. Half a rotation equals 180 degrees or π radians and represents a **half-circle**. Subsequently, 90 degrees ($\pi/2$ radians) represents a quarter of a circle, which is known as a **right angle**. Any angle less than 90 degrees is an **acute angle**, and any angle greater than 90 degrees is an **obtuse angle**. Angle measurement is additive. When an angle is broken into two non-overlapping angles, the total measure of the larger angle equals the sum of the two smaller angles. Lines are *coplanar* if they're located in the same plane. Two lines are **parallel** if they are coplanar, extend in the same direction, and never cross. If lines do cross, they're labeled as **intersecting lines** because they "intersect" at one point. If they intersect at more than one point, they're the same line. **Perpendicular lines** are coplanar lines that form a right angle at their point of intersection.

Transformations in the Plane

A **transformation** occurs when a shape is altered in the plane where it exists. There are three major types of transformation: translations, reflections, and rotations. A **translation** consists of shifting a shape in one direction. A **reflection** results when a shape is transformed over a line to its mirror image. Finally, a **rotation** occurs when a shape moves in a circular motion around a specified point. The object can be turned clockwise or counterclockwise and, if rotated 360 degrees, returns to its original location.

Transformations as Functions

A **function** is when a translation, reflection, or rotation (i.e., transformation) occurs in the coordinate plane. The original points are the **inputs** of the function, and the resulting points are the **outputs** of the function. For instance, if a shape is shifted to the right 4 units in the coordinate plane, the original x variable becomes $x + 4$. If a shape is reflected over the y-axis, all x coordinates are negated. For instance, if the original shape reaches the line $x = 4$, the resulting shape would reach the line $x = -4$.

Distance and Angle Measure

The three major types of transformations preserve distance and angle measurement. The shapes stay the same, but they are moved to another place in the plane. Therefore, the distance between any two points on the shape doesn't change. Also, any original angle measure between two line segments doesn't change. However, there are transformations that don't preserve distance and angle measurements, including those that don't preserve the original shape. For example, transformations that involve stretching and shrinking shapes don't preserve distance and angle measures. In these cases, the input variables are multiplied by either a number greater than one (**stretch**) or less than one (**shrink**).

Rotations and Reflections

A **point of symmetry** is used to determine translations that map a shape onto itself. When a line is drawn through a point of symmetry, it crosses the shape on one side of the point while crossing the shape on the other side at the exact same distance. A shape can be rotated 180 degrees around a point of symmetry to get back to its original shape. Simple examples are the center of a circle and a square. A **line of symmetry** is a line that a shape is folded over to have its sides align, and it goes through the point of symmetry. A combination of reflecting an original shape around a line of symmetry and rotating it can map the original image onto itself. A shape has **rotational symmetry** if it can be rotated to get back to its original shape, and it has **reflection symmetry** if it can be reflected onto itself over some line. Shapes such as rectangles, parallelograms, trapezoids, and regular polygons can be mapped onto themselves through such rotations and reflections.

Isometric Transformations

Rotations, reflections, and translations are isometric transformations, because throughout each transformation the distance of line segments is maintained, the angle measure is maintained, parallel lines in the original shape remain parallel, and points on lines remain on those lines. A rotation turns a shape around a specific point (O) known as the **center of rotation**. An **angle of rotation** is formed by drawing a ray from the center of rotation to a point (P) on the original shape and to the point's image (P') on the reflected shape. Thus, it's true that $OP=OP'$. A reflection over a line (l), known as the **line of reflection**, takes an original point P and maps it to its image P' on the opposite side of l. The line of reflection is the perpendicular bisector of every line formed by an original point and its image. A translation maps each point P in the original shape to a new point P'. The line segment formed between each point and its image consists of the same length, and the line segment formed by two original points is parallel to the line segment formed from their two images.

Transformation Figures

Once the major transformations are defined, a shape can be altered by performing given transformations. An image is determined from a pre-image by carrying out a series of rotations, reflections, and translations. Once transformations are understood completely, moving and manipulating figures can be related to real-word situations. The transformed figure can then be drawn using either a pencil and paper or geometry software.

Transformation Mapping

Given a pre-image and an image, transformations can be determined that turn one shape into the other. A sequence of rotations, reflections, and translations can be defined that map the pre-image onto the other shape.

Proving Geometric Theorems

Proving Theorems About Lines and Angles

To prove any geometric theorem, the proven theorems must be linked in a logical order that flows from an original point to the desired result. Proving theorems about lines and angles is the basis of proving theorems that involve other shapes. A **transversal** is a line that passes through two lines at two points. Common theorems that need to be proved are: vertical angles are congruent; a transversal passing through two parallel lines forms two alternate interior angles that are congruent and two corresponding angles that are congruent; and points on a perpendicular bisector of a line segment are equidistant from the endpoints of the line segment.

Triangle Theorems

To prove theorems about triangles, basic definitions involving triangles (e.g., equilateral, isosceles, etc.) need to be known. Proven theorems concerning lines and angles can be applied to prove theorems about triangles. Common theorems to be proved include: the sum of all angles in a triangle equals 180 degrees; the sum of the lengths of two sides of a triangle is greater than the length of the third side; the base angles of an isosceles triangle are congruent; the line segment connecting the midpoint of two sides of a triangle is parallel to the third side and its length is half the length of the third side; and the medians of a triangle all meet at a single point.

Parallelogram Theorems

A **parallelogram** is a quadrilateral with parallel opposing sides. Within parallelograms, opposite sides and angles are congruent and the diagonals bisect each other. Known theorems about parallel lines, transversals, complementary angles, and congruent triangles can be used to prove theorems about parallelograms. Theorems that need to be proved include: opposite sides of a parallelogram are congruent; opposite angles are congruent; the diagonals bisect each other; and rectangles are parallelograms with congruent diagonals.

Geometric Constructions Made with a Variety of Tools and Methods

Geometric Construction Tools

The tools needed to make formal geometric constructions are a compass, a ruler, paper folding, or geometry software. These tools can be used to copy or bisect a line segment, bisect an angle, construct perpendicular lines, construct a perpendicular bisector of a line segment, and construct a line parallel to a given line through a specified point.

Formal Geometric Constructions

Beginning with formal geometric constructions, various geometric figures and shapes can be built. Definitions and theorems for lines and angles can be used in parallel with geometric constructions to build shapes such as equilateral triangles, squares, and rectangular hexagons inscribed in a circle. Definitions of shapes involving congruence of sides and angles within each type of figure must be understood and used in parallel with constructing congruent line segments, parallel and perpendicular lines, and congruent angles.

Congruence and Similarity in Terms of Transformations

Rigid Motion

A **rigid motion** is a transformation that preserves distance and length. Every line segment in the resulting image is congruent to the corresponding line segment in the pre-image. **Congruence** between two figures means a series of transformations (or a rigid motion) can be defined that maps one of the figures onto the other. Basically, two figures are congruent if they have the same shape and size.

Dilation

A shape is dilated, or a **dilation** occurs, when each side of the original image is multiplied by a given **scale factor**. If the scale factor is less than 1 and greater than 0, the dilation contracts the shape, and the resulting shape is smaller. If the scale factor equals 1, the resulting shape is the same size, and the dilation is a rigid motion. Finally, if the scale factor is greater than 1, the resulting shape is larger and the dilation expands the shape. The **center of dilation** is the point where the distance from it to any point on the new shape equals the scale factor multiplied by the distance from the center to the corresponding point in the pre-image. Dilation isn't an isometric transformation because distance isn't preserved.

However, angle measure, parallel lines, and points on a line all remain unchanged. The following figure is an example of translation, rotation, dilation, and reflection:

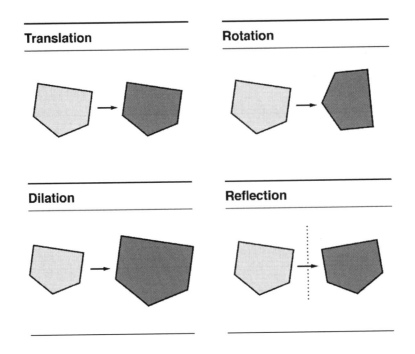

Translation

Rotation

Dilation

Reflection

Determining Congruence
Two figures are congruent if there is a rigid motion that can map one figure onto the other. Therefore, all pairs of sides and angles within the image and pre-image must be congruent. For example, in triangles, each pair of the three sides and three angles must be congruent. Similarly, in two four-sided figures, each pair of the four sides and four angles must be congruent.

Similarity
Two figures are **similar** if there is a combination of translations, reflections, rotations, and dilations, which maps one figure onto the other. The difference between congruence and similarity is that dilation can be used in similarity. Therefore, side lengths between each shape can differ. However, angle measure must be preserved within this definition. If two polygons differ in size so that the lengths of corresponding line segments differ by the same factor, but corresponding angles have the same measurement, they are similar.

Triangle Congruence
There are five theorems to show that triangles are congruent when it's unknown whether each pair of angles and sides are congruent. Each theorem is a shortcut that involves different combinations of sides and angles that must be true for the two triangles to be congruent. For example, **side-side-side (SSS)** states that if all sides are equal, the triangles are congruent. **Side-angle-side (SAS)** states that if two pairs of sides are equal and the included angles are congruent, then the triangles are congruent. Similarly, **angle-side-angle (ASA)** states that if two pairs of angles are congruent and the included side lengths are equal, the triangles are similar. **Angle-angle-side (AAS)** states that two triangles are congruent if they have two pairs of congruent angles and a pair of corresponding equal side lengths that aren't included. Finally, **hypotenuse-leg (HL)** states that if two right triangles have equal hypotenuses and an equal pair of shorter sides, then the triangles are congruent. An important item to note is that angle-angle-angle *(AAA)* is not enough information to have congruence. It's important to understand

why these rules work by using rigid motions to show congruence between the triangles with the given properties. For example, three reflections are needed to show why *SAS* follows from the definition of congruence.

Similarity for Two Triangles

If two angles of one triangle are congruent with two angles of a second triangle, the triangles are similar. This is because, within any triangle, the sum of the angle measurements is 180 degrees. Therefore, if two are congruent, the third angle must also be congruent because their measurements are equal. Three congruent pairs of angles mean that the triangles are similar.

Proving Congruence and Similarity

The criteria needed to prove triangles are congruent involves both angle and side congruence. Both pairs of related angles and sides need to be of the same measurement to use congruence in a proof. The criteria to prove similarity in triangles involves proportionality of side lengths. Angles must be congruent in similar triangles; however, corresponding side lengths only need to be a constant multiple of each other. Once similarity is established, it can be used in proofs as well. Relationships in geometric figures other than triangles can be proven using triangle congruence and similarity. If a similar or congruent triangle can be found within another type of geometric figure, their criteria can be used to prove a relationship about a given formula. For instance, a rectangle can be broken up into two congruent triangles.

Trigonometric Ratios in Right Triangles

Trigonometric Functions

Within similar triangles, corresponding sides are proportional, and angles are congruent. In addition, within similar triangles, the ratio of the side lengths is the same. This property is true even if side lengths are different. Within right triangles, trigonometric ratios can be defined for the acute angle within the triangle. The functions are defined through ratios in a right triangle. Sine of acute angle, A, is opposite over hypotenuse, cosine is adjacent over hypotenuse, and tangent is opposite over adjacent. Note that expanding or shrinking the triangle won't change the ratios. However, changing the angle measurements will alter the calculations.

Complementary Angles

Angles that add up to 90 degrees are **complementary.** Within a right triangle, two complementary angles exist because the third angle is always 90 degrees. In this scenario, the **sine** of one of the complementary angles is equal to the **cosine** of the other angle. The opposite is also true. This relationship exists because sine and cosine will be calculated as the ratios of the same side lengths.

The Pythagorean Theorem

The **Pythagorean theorem** is an important relationship between the three sides of a right triangle. It states that the square of the side opposite the right triangle, known as the **hypotenuse** (denoted as c^2), is equal to the sum of the squares of the other two sides ($a^2 + b^2$). Thus, $a^2 + b^2 = c^2$.

Both the trigonometric functions and the Pythagorean theorem can be used in problems that involve finding either a missing side or a missing angle of a right triangle. To do so, one must look to see what sides and angles are given and select the correct relationship that will help find the missing value. These relationships can also be used to solve application problems involving right triangles. Often, it's helpful to draw a figure to represent the problem to see what's missing.

Application of Trigonometry to General Triangles

The Area Formula
A triangle that isn't a right triangle is known as an **oblique triangle**. It should be noted that even if the triangle consists of three acute angles, it is still referred to as an oblique triangle. **Oblique,** in this case, does not refer to an angle measurement. Consider the following oblique triangle:

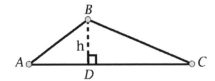

For this triangle, $Area = \frac{1}{2} \times base \times height = \frac{1}{2} \times AC \times BD$. The **auxiliary line** drawn from the vertex B perpendicular to the opposite side AC represents the height of the triangle. This line splits the larger triangle into two smaller right triangles, which allows for the use of the trigonometric functions (specifically that $\sin A = \frac{h}{AB}$). Therefore, $Area = \frac{1}{2} \times \boldsymbol{AC} \times \boldsymbol{AB} \times \sin A$. Typically the sides are labelled as the lowercase letter of the vertex that's opposite. Therefore, the formula can be written as $Area = \frac{1}{2} ab \sin A$. This area formula can be used to find areas of triangles when given side lengths and angle measurements, or it can be used to find side lengths or angle measurements based on a specific area and other characteristics of the triangle.

Laws of Sines and Cosines
The **law of sines** and **law of cosines** are two more relationships that exist within oblique triangles. Consider a triangle with sides a, b, and c, and angles A, B, and C opposite the corresponding sides.

The law of cosines states that $c^2 = a^2 + b^2 - 2ab \cos C$. The law of sines states that $\frac{\sin A}{a} = \frac{\sin B}{b} = \frac{\sin C}{c}$. In addition to the area formula, these two relationships can help find unknown angle and side measurements in oblique triangles.

Circle Theorems

Circle Angles
The **radius** of a circle is the distance from the center of the circle to any point on the circle. A **chord** of a circle is a straight line formed when its endpoints are allowed to be any two points on the circle. Many angles exist within a circle. A **central angle** is formed by using two radii as its rays and the center of the circle as its vertex. An **inscribed angle** is formed by using two chords as its rays, and its vertex is a point on the circle itself. Finally, a **circumscribed angle** has a vertex that is a point outside the circle and rays that intersect with the circle. Some relationships exist between these types of angles, and, in order to define these relationships, arc measure must be understood. An **arc** of a circle is a portion of the circumference. Finding the **arc measure** is the same as finding the degree measure of the central angle that intersects the circle to form the arc. The measure of an inscribed angle is half the measure of its intercepted arc. It's also true that the measure of a circumscribed angle is equal to 180 degrees minus the measure of the central angle that forms the arc in the angle.

Quadrilateral Angles

If a quadrilateral is inscribed in a circle, the sum of its opposite angles is 180 degrees. Consider the quadrilateral ABCD centered at the point O:

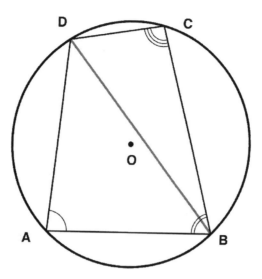

Each of the four line segments within the quadrilateral is a chord of the circle. Consider the diagonal DB. Angle DAB is an inscribed angle leaning on the arc DCB. Therefore, angle DAB is half the measure of the arc DCB. Conversely, angle DCB is an inscribed angle leaning on the arc DAB. Therefore, angle DCB is half the measure of the arc DAB. The sum of arcs DCB and DAB is 360 degrees because they make up the entire circle. Therefore, the sum of angles DAB and DCB equals half of 360 degrees, which is 180 degrees.

Circle Lines

A **tangent line** is a line that touches a curve at a single point without going through it. A **compass** and a **straight edge** are the tools necessary to construct a tangent line from a point P outside the circle to the circle. A tangent line is constructed by drawing a line segment from the center of the circle O to the point P, and then finding its midpoint M by bisecting the line segment. By using M as the center, a **compass** is used to draw a circle through points O and P. N is defined as the intersection of the two circles. Finally, a line segment is drawn through P and N. This is the **tangent line**. Each point on a circle has only one tangent line, which is perpendicular to the radius at that point. A line similar to a tangent line is a **secant line.** Instead of intersecting the circle at one point, a secant line intersects the circle at two points. A **chord** is a smaller portion of a secant line.

Here's an example:

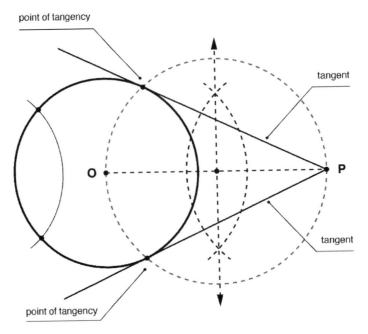

point of tangency

tangent

O

P

tangent

point of tangency

Arc Length and Area Measurements of Sectors of Circles

Arc Length

As previously mentioned, angles can be measured in radians, and 180 degrees equals π radians. Therefore, the measure of a complete circle is 2π radians. In addition to arc measure, **arc length** can also be found because the length of an arc is a portion of the circle's circumference. The following proportion is true:

$$\frac{\text{Arc measure}}{360 \text{ degrees}} = \frac{\text{arc length}}{\text{arc circumference}}$$

Arc measure is the same as the measure of the central angle, and this proportion can be rewritten as:

$$\text{arc length} = \frac{\text{central angle}}{360 \text{ degrees}} \times \text{circumference}$$

In addition, the degree measure can be replaced with radians to allow the use of both units. Note that arc length is a fractional part of circumference because $\frac{\text{central angle}}{360 \text{ degrees}} < 1$.

Area of a Sector

A **sector** of a circle is a portion of the circle that's enclosed by two radii and an arc. It resembles a piece of a pie, and the area of a sector can be derived using known definitions. The area of a circle can be calculated using the formula $A = \pi r^2$, where r is the radius of the circle. The area of a sector of a circle is a fraction of that calculation. For example, if the central angle θ is known in radians, the area of a sector is defined as:

$$A_s = \pi r^2 \frac{\theta}{2\pi} = \frac{\vartheta r^2}{2}$$

82

If the angle θ in degrees is known, the area of the sector is $A_s = \frac{\vartheta \pi r^2}{360}$. Finally, if the arc length L is known, the area of the sector can be reduced to $A_s = \frac{rL}{2}$.

Translating Between a Geometric Description and an Equation for a Conic Section

<u>Equation of a Circle</u>
A **circle** can be defined as the set of all points that are the same distance (known as the **radius**, r) from a single point C (known as the center of the circle). The center has coordinates (h, k), and any point on the circle can be labelled with coordinates (x, y).

As shown below, a **right triangle** is formed with these two points:

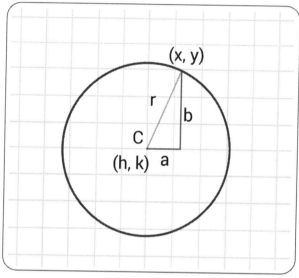

The **Pythagorean theorem** states that $a^2 + b^2 = r^2$. However, a can be replaced by $|x - h|$ and b can be replaced by $|y - k|$ by using the **distance formula** which is:

$$d = \sqrt{(x_2 - x_1)^2 + (y_2 - y_1)^2}$$

That substitution results in:

$$(x - h)^2 + (y - k)^2 = r^2$$

This is the formula for finding the equation of any circle with a center (h, k) and a radius r. Note that sometimes c is used instead of r.

<u>Finding the Center and Radius</u>
Circles aren't always given in the form of the circle equation where the center and radius can be seen so easily. Oftentimes, they're given in the more general format of $ax^2 + by^2 + cx + dy + e = 0$. This can be converted to the center-radius form using the algebra technique of completing the square in both variables. First, the constant term is moved over to the other side of the equals sign, and then the x and y variable terms are grouped together. Then the equation is divided through by a and, because this is the equation of a circle, $a = b$. At this point, the x-term coefficient is divided by 2, squared, and then added to both sides of the equation. This value is grouped with the x terms. The same steps then need to be completed with the y-term coefficient. The trinomial in both x and y can now be factored into a square of a binomial, which gives both $(x - h)^2$ and $(y - k)^2$.

Parabola Equations

A **parabola** is defined as a specific type of curve such that any point on it is the same distance from a fixed point (called the **foci**) and a fixed straight line (called the **directrix**). A parabola is the shape formed from the intersection of a cone with a plane that's parallel to its side. Every parabola has an **axis of symmetry**, and its **vertex** (h, k) is the point at which the axis of symmetry intersects the curve. If the parabola has an axis of symmetry parallel to the y-axis, the focus is the point $(h, k + f)$ and the directrix is the line $y = k - f$. For example, a parabola may have a vertex at the origin, focus $(0, f)$, and directrix $y = -f$. The equation of this parabola can be derived by using both the focus and the directrix. The distance from any coordinate on the curve to the focus is the same as the distance to the directrix, and the Pythagorean theorem can be used to find the length of d. The triangle has sides with length $|x|$ and $|y - f|$ and therefore:

$$d = \sqrt{x^2 + (y - f)^2}$$

By definition, the **vertex** is halfway between the focus and the directrix and $d = y + f$. Setting these two equations equal to one another, squaring each side, simplifying, and solving for y gives the equation of a parabola with the focus f and the vertex being the origin $y = \frac{1}{4f}x^2$. If the vertex (h, k) is not the origin, a similar process can be completed to derive the equation $(x - h)^2 = 4f(y - k)$ for a parabola with focus f.

Ellipse and Hyperbola Equations

An **ellipse** is the set of all points for which the sum of the distances from two fixed points (known as the *foci*) is constant. A **hyperbola** is the set of all points for which the difference between the distances from two fixed points (also known as the *foci*) is constant. The **distance formula** can be used to derive the formulas of both an ellipse and a hyperbola, given the coordinates of the foci. Consider an ellipse where its major axis is horizontal (i.e., it's longer along the x-axis) and its foci are the coordinates $(-c, 0)$ and $(c, 0)$. The distance from any point (x, y) to $(-c, 0)$ is $d_1 = \sqrt{(x + c)^2 + y^2}$, and the distance from the same point (x, y) to $(c, 0)$ is:

$$d_1 = \sqrt{(x - c)^2 + y^2}$$

Using the definition of an ellipse, it's true that the sum of the distances from the vertex a to each foci is equal to $d_1 + d_2$. Therefore:

$$d_1 + d_2 = (a + c) + (a - c) = 2a$$

and

$$\sqrt{(x + c)^2 + y^2} + \sqrt{(x - c)^2 + y^2} = 2a$$

After a series of algebraic steps, this equation can be simplified to $\frac{x^2}{a^2} + \frac{y^2}{b^2} = 1$, which is the equation of an ellipse with a horizontal major axis. In this case, $a > b$. When the ellipse has a vertical major axis, similar techniques result in $\frac{x^2}{b^2} + \frac{y^2}{a^2} = 1$, and $a > b$.

The equation of a hyperbola can be derived in a similar fashion. Consider a hyperbola with a horizontal major axis and its foci are also the coordinates $(-c, 0)$ and $(c, 0)$. Again, the distance from any point (x, y) to $(-c, 0)$ is $d_1 = \sqrt{(x + c)^2 + y^2}$ and the distance from the same point (x, y) to $(c, 0)$ is:

$$d_1 = \sqrt{(x - c)^2 + y^2}$$

Using the definition of a hyperbola, it's true that the difference of the distances from the vertex a to each foci is equal to $d_1 - d_2$. Therefore:

$$d_1 - d_2 = (c + a) - (c - a) = 2a$$

This means that:

$$\sqrt{(x + c)^2 + y^2} - \sqrt{(x - c)^2 + y^2} = 2a$$

After a series of algebraic steps, this equation can be simplified to:

$$\frac{x^2}{a^2} - \frac{y^2}{b^2} = 1$$

This is the equation of a hyperbola with a horizontal major axis. In this case, $a > b$. Similar techniques result in the equation $\frac{x^2}{b} - \frac{y^2}{a^2} = 1$, where $a > b$, when the hyperbola has a vertical major axis.

Using Coordinate Geometry to Algebraically Prove Simple Geometric Theorems

Proving Theorems with Coordinates

Many important formulas and equations exist in geometry that use coordinates. The distance between two points (x_1, y_1) and (x_2, y_2) is:

$$d = \sqrt{(x_2 - x_1)^2 + (y_2 - y_1)^2}.$$

The slope of the line containing the same two points is $m = \frac{y_2 - y_1}{x_2 - x_1}$. Also, the midpoint of the line segment with endpoints (x_1, y_1) and (x_2, y_2) is:

$$M = \left(\frac{x_1 + x_2}{2}, \frac{y_1 + y_2}{2}\right)$$

The equations of a circle, parabola, ellipse, and hyperbola can also be used to prove theorems algebraically. Knowing when to use which formula or equation is extremely important, and knowing which formula applies to which property of a given geometric shape is an integral part of the process. In some cases, there are a number of ways to prove a theorem; however, only one way is required.

Solving Problems with Parallel and Perpendicular Lines

Two lines can be parallel, perpendicular, or neither. If two lines are **parallel**, they have the same slope. This is proven using the idea of similar triangles. Consider the following diagram with two parallel lines, L1 and L2:

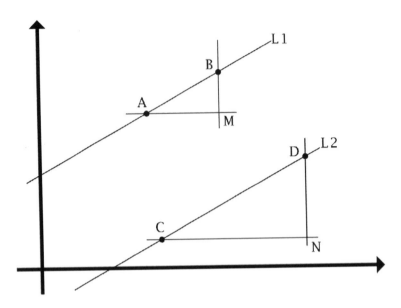

A and B are points on L1, and C and D are points on L2. Right triangles are formed with vertex M and N where lines BM and DN are parallel to the y-axis and AM and CN are parallel to the x-axis. Because all three sets of lines are parallel, the triangles are similar. Therefore, $\frac{BM}{DN} = \frac{MA}{NC}$. This shows that the rise/run is equal for lines L1 and L2. Hence, their slopes are equal.

Secondly, if two lines are **perpendicular**, the product of their slopes equals -1. This means that their slopes are negative reciprocals of each other. Consider two perpendicular lines, *l* and *n*:

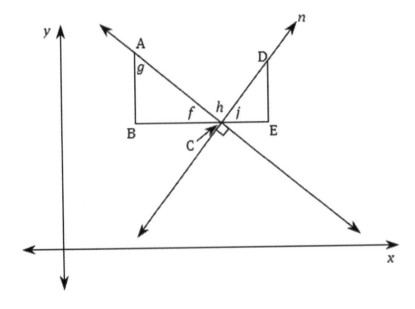

86

Right triangles ABC and CDE are formed so that lines BC and CE are parallel to the x-axis, and AB and DE are parallel to the y-axis. Because line BE is a straight line, angles $f + h + i = 180\ degrees$. However, angle h is a right angle, so $f + j = 90\ degrees$. By construction, $f + g = 90$, which means that $g = j$. Therefore, because angles $B = E$ and $g = j$, the triangles are similar and $\frac{AB}{BC} = \frac{CE}{DE}$. Because slope is equal to rise/run, the slope of line l is $-\frac{AB}{BC}$ and the slope of line n is $\frac{DE}{CE}$. Multiplying the slopes together gives:

$$-\frac{AB}{BC} \cdot \frac{DE}{CE} = -\frac{CE}{DE} \cdot \frac{DE}{CE} = -1$$

This proves that the product of the slopes of two perpendicular lines equals -1. Both parallel and perpendicular lines can be integral in many geometric proofs, so knowing and understanding their properties is crucial for problem-solving.

Formulas for Ratios
If a line segment with endpoints (x_1, y_1) and (x_2, y_2) is partitioned into two equal parts, the formula for **midpoint** is used. Recall this formula is:

$$M = \left(\frac{x_1 + x_2}{2}, \frac{y_1 + y_2}{2}\right)$$

The ratio of line segments is 1:1. However, if the ratio needs to be anything other than 1:1, a different formula must be used. Consider a ratio that is $a:b$. This means the desired point that partitions the line segment is $\frac{a}{a+b}$ of the way from (x_1, y_1) to (x_2, y_2). The actual formula for the coordinate is:

$$\left(\frac{bx_1 + ax_2}{a + b}, \frac{by_1 + ay_2}{a + b}\right)$$

Computing Side Length, Perimeter, and Area
The side lengths of each shape can be found by plugging the endpoints into the distance formula between two ordered pairs (x_1, y_1) and (x_2, y_2).

As a reminder, this is the **distance formula**:

$$d = \sqrt{(x_2 - x_1)^2 + (y_2 - y_1)^2}$$

The distance formula is derived from the Pythagorean theorem. Once the side lengths are found, they can be added together to obtain the perimeter of the given polygon. Simplifications can be made for specific shapes such as squares and equilateral triangles. For example, one side length can be multiplied by 4 to obtain the perimeter of a square. Also, one side length can be multiplied by 3 to obtain the perimeter of an equilateral triangle. A similar technique can be used to calculate areas. For polygons, both side length and height can be found by using the same distance formula. Areas of triangles and quadrilaterals are straightforward through the use of $A = \frac{1}{2}bh$ or $A = bh$, depending on the shape. To find the area of other polygons, their shapes can be partitioned into rectangles and triangles. The areas of these simpler shapes can be calculated and then added together to find the total area of the polygon.

Perimeter, Area, Surface Area, and Volume Formulas

Circumference and Volume

The **circumference** of a circle is found by calculating the perimeter of the shape. The ratio of the circumference to the diameter of the circle is π, so the formula for circumference is $C = \pi d = 2\pi r$. To visualize this, one can imagine that a circle is a pie divided into an equal number of slices. The slices can be aligned to form a parallelogram with a height equal to the radius r, and a base equal to half of the circumference of the circle πr. Plugging these expressions into the formula for area of a parallelogram results in $A = bh = \pi r^2$. The **volume** of a cylinder is then found by adding a third dimension onto the circle. Volume of a cylinder is calculated by multiplying the area of the base (which is a circle) by the height of the cylinder. Doing so results in the equation $V = \pi r^2 h$. Next, consider the volume of a rectangular box $= lwh$, where l is length, w is width, and h is height. This can be simplified into $V = Ah$, where A is the area of the base. The volume of a pyramid with the same dimensions is $1/3$ of this quantity because it fills up $1/3$ of the space. Therefore, the volume of a pyramid is $V = {}^1\!/_3\, Ah$. In a similar fashion, the volume of a cone is ${}^1\!/_3$ of the volume of a cylinder. Therefore, the formula for the volume of a cone is ${}^1\!/_3\, \pi r^2 h$.

Perimeter and Area

Both the perimeter and area formulas are applicable in real-world scenarios. Knowing the **perimeter** is useful when the length of a shape's outline is needed. For example, to build a fence around a yard, the yard's perimeter must be calculated so enough materials are purchased to complete the fence. The **area** is necessary anytime the surface of a shape is needed. For example, when constructing a garden, the area of the garden region is needed so enough dirt can be purchased to fill it. Many times, it's necessary to break up the given shape into shapes with known perimeter and area formulas (such as triangles and rectangles) and add the individual perimeters or areas together to determine the desired quantity.

Surface Area and Volume

Many real-world objects are a combination of prisms, cylinders, pyramids, and spheres. **Surface area** problems relate to quantifying the outside area of such a three-dimensional object, and **volume** problems involve quantifying how much the three-dimensional object can hold. For example, when calculating how much paint is needed to paint an entire house, surface area is used. Conversely, when calculating how much water a cylindrical tank can hold, volume is used. The surface area of a **prism** is the sum of all the areas, which simplifies into $SA = 2A + Bh$ where A is the area of the base, B is the perimeter of the base, and h is the height of the prism. The volume of the same prism is $V = Ah$. The surface area of a **cylinder** is the sum of the areas of both ends and the side, which is $SA = 2\pi rh + 2\pi r^2$. The surface area of a **pyramid** is calculated by adding known area formulas. It is equal to the area of the base (which is rectangular) plus the area of the four triangles that form the sides. The surface area of a **cone** is equal to the area of the base plus the area of the top, which is:

$$ SA = \pi r^2 + \pi \pi r \sqrt{h^2 + r^2} $$

Finally, the surface area of a **sphere** is $SA = 4\pi r^2$ and its volume is $V = \dfrac{4}{3}\pi r^3$.

Visualizing Relationships Between Two-Dimensional and Three-Dimensional Objects

Cross Sections and Rotations

Two-dimensional objects are formed when three-dimensional objects are "sliced" in various ways. For example, any cross-section of a sphere is a circle. Some three-dimensional objects have different cross sections depending on how the object is sliced. For example, the cross section of a cylinder can be a circle or a rectangle, and the cross section of a pyramid can be a square or a triangle. In addition, three-dimensional objects can be formed by rotating two-dimensional objects. Certain rotations can relate the two-dimensional cross sections back to the original three-dimensional objects. The objects must be rotated around an imaginary line known as the **rotation axis.** For example, a right triangle can be rotated around one of its legs to form a cone. A sphere can be formed by rotating a semicircle around a line segment formed from its diameter. Finally, rotating a square around one of its sides forms a cylinder.

Simplifying Three-Dimensional Objects

Three-dimensional objects can be simplified into related two-dimensional shapes to solve problems. This simplification can make problem-solving a much easier experience. An isometric representation of a three-dimensional object can be completed so that important properties (e.g., shape, relationships of faces and surfaces) are noted. Edges and vertices can be translated into two-dimensional objects as well. For example, below is a three-dimensional object that's been partitioned into two-dimensional representations of its faces:

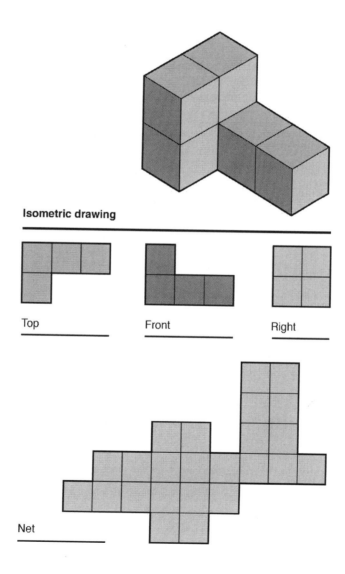

Isometric drawing

Top   Front   Right

Net

89

The **net** represents the sum of the three different faces. Depending on the problem, using a smaller portion of the given shape may be helpful, by simplifying the steps necessary to solve.

Applying Geometric Concepts to Real-World Situations

Real-World Geometry

Many real-world objects can be compared to geometric shapes. Describing certain objects using the measurements and properties of two- and three-dimensional shapes is an important part of geometry. For example, basic ideas such as angles and line segments can be seen in real-world objects. The corner of any room is an angle, and the intersection of a wall with the floor is like a line segment. Building upon this idea, entire objects can be related to both two- and three-dimensional shapes. An entire room can be thought of as square, rectangle, or a sum of a few three-dimensional shapes. Knowing what properties and measures are needed to make decisions in real life is why geometry is such a useful branch of mathematics. One obvious relationship between a real-life situation and geometry exists in construction. For example, to build an addition onto a house, several geometric measurements will be used.

Density

The **density** of a substance is the ratio of mass to area or volume. It's a relationship between the mass and how much space the object actually takes up. Knowing which units to use in each situation is crucial. Population density is an example of a real-life situation that's modeled by using density concepts. It involves calculating the ratio of the number of people to the number of square miles. The amount of material needed per a specific unit of area or volume is another application. For example, estimating the number of BTUs per cubic foot of a home is a measurement that relates to heating or cooling the house based on the desired temperature and the house's size.

Solving Design Problem

Design problems are an important application of geometry (e.g., building structures that satisfy physical constraints and/or minimize costs). These problems involve optimizing a situation based on what's given and required. For example, determining what size barn to build, given certain dimensions and a specific budget, uses both geometric properties and other mathematical concepts. Equations are formed using geometric definitions and the given constraints. In the end, such problems involve solving a system of equations and rely heavily on a strong background in algebra. **Typographic grid systems** also help with such design problems. A grid made up of intersecting straight or curved lines can be used as a visual representation of the structure being designed. This concept is seen in the blueprints used throughout the graphic design process.

Properties of Parallel and Perpendicular Lines, Triangles, Quadrilaterals, Polygons, and Circles

Solving Line Problems

Two lines are parallel if they have the same slope and different intercept. Two lines are perpendicular if the product of their slope equals -1. Parallel lines never intersect unless they are the same line, and perpendicular lines intersect at a right angle. If two lines aren't parallel, they must intersect at one point. Determining equations of lines based on properties of parallel and perpendicular lines appears in word problems. To find an equation of a line, both the slope and a point the line goes through are necessary. Therefore, if an equation of a line is needed that's parallel to a given line and runs through a specified point, the slope of the given line and the point are plugged into the point-slope form of an equation of a line. Secondly, if an equation of a line is needed that's perpendicular to a given line running through a specified point, the negative reciprocal of the slope of the given line and the point are plugged into the **point-slope form**. Also, if the point of intersection of two lines is known, that point will be used to solve the set of equations. Therefore, to solve a system of equations, the point of intersection must be found. If a set of two equations with two unknown variables has no solution, the lines are parallel.

Relationships between Angles

Supplementary angles add up to 180 degrees. **Vertical angles** are two nonadjacent angles formed by two intersecting lines. **Corresponding angles** are two angles in the same position whenever a straight line (known as a **transversal**) crosses two others. If the two lines are parallel, the corresponding angles are equal. **Alternate interior angles** are also a pair of angles formed when two lines are crossed by a transversal. They are opposite angles that exist inside of the two lines. In the corresponding angles diagram above, angles 2 and 7 are alternate interior angles, as well as angles 6 and 3. **Alternate exterior angles** are opposite angles formed by a transversal but, in contrast to interior angles, exterior angles exist outside the two original lines. Therefore, angles 1 and 8 are alternate exterior angles and so are angles 5 and 4. Finally, **consecutive interior angles** are pairs of angles formed by a transversal. These angles are located on the same side of the transversal and inside the two original lines. Therefore, angles 2 and 3 are a pair of consecutive interior angles, and so are angles 6 and 7. These definitions are instrumental in solving many problems that involve determining relationships between angles.

Medians, Midpoints, and Altitudes

A **median** of a triangle is the line drawn from a vertex to the midpoint on the opposite side. A triangle has three medians, and their point of intersection is known as the **centroid**. An **altitude** is a line drawn from a vertex perpendicular to the opposite side. A triangle has three altitudes, and their point of intersection is known as the **orthocenter**. An altitude can actually exist outside, inside, or on the triangle depending on the placement of the vertex. Many problems involve these definitions. For example, given one endpoint of a line segment and the midpoint, the other endpoint can be determined by using the midpoint formula. In addition, area problems heavily depend on these definitions. For example, it can be proven that the median of a triangle divides it into two regions of equal areas. The actual formula for the area of a triangle depends on its altitude.

Special Triangles

An **isosceles triangle** contains at least two equal sides. Therefore, it must also contain two equal angles and, subsequently, contain two medians of the same length. An isosceles triangle can also be labelled as an **equilateral triangle** (which contains three equal sides and three equal angles) when it meets these conditions. In an equilateral triangle, the measure of each angle is always 60 degrees. Also within an equilateral triangle, the medians are of the same length. A **scalene triangle** can never be an equilateral or an isosceles triangle because it contains no equal sides and no equal angles. Also, medians in a scalene triangle can't have the same length. However, a **right triangle**, which is a triangle containing a 90-degree angle, can be a scalene triangle. There are two types of special right triangles. The **30-60-90 right triangle** has angle measurements of 30 degrees, 60 degrees, and 90 degrees. Because of the nature of this triangle, and through the use of the Pythagorean theorem, the side lengths have a special relationship. If x is the length opposite the 30-degree angle, the length opposite the 60-degree angle is $\sqrt{3}x$, and the hypotenuse has length $2x$. The **45-45-90 right triangle** is also special as it contains two angle measurements of 45 degrees. It can be proven that, if x is the length of the two equal sides, the hypotenuse is $x\sqrt{2}$. The properties of all of these special triangles are extremely useful in determining both side lengths and angle measurements in problems where some of these quantities are given and some are not.

Special Quadrilaterals

A special quadrilateral is one in which both pairs of opposite sides are parallel. This type of quadrilateral is known as a **parallelogram**. A parallelogram has six important properties:

- Opposite sides are congruent.
- Opposite angles are congruent.
- Within a parallelogram, consecutive angles are supplementary, so their measurements total 180 degrees.
- If one angle is a right angle, all of them have to be right angles.
- The diagonals of the angles bisect each other.
- These diagonals form two congruent triangles.

A parallelogram with four congruent sides is a **rhombus**. A quadrilateral containing only one set of parallel sides is known as a **trapezoid**. The parallel sides are known as **bases**, and the other two sides are known as **legs**. If the legs are congruent, the trapezoid can be labelled an **isosceles trapezoid**. An important property of a trapezoid is that their **diagonals** are congruent. Also, the **median** of a trapezoid is parallel to the bases, and its length is equal to half of the sum of the base lengths.

Quadrilateral Relationships

Rectangles, squares, and rhombuses are **polygons** with four sides. By definition, all rectangles are parallelograms, but only some rectangles are squares. However, some parallelograms are rectangles. Also, it's true that all squares are rectangles, and some rhombuses are squares. There are no rectangles, squares, or rhombuses that are trapezoids though, because they have more than one set of parallel sides.

Diagonals and Angles

Diagonals are lines (excluding sides) that connect two vertices within a polygon. **Mutually bisecting diagonals** intersect at their midpoints. Parallelograms, rectangles, squares, and rhombuses have mutually bisecting diagonals. However, trapezoids don't have such lines. **Perpendicular diagonals** occur when they form four right triangles at their point of intersection. Squares and rhombuses have perpendicular diagonals, but trapezoids, rectangles, and parallelograms do not. Finally, **perpendicular bisecting diagonals** (also known as **perpendicular bisectors**) form four right triangles at their point of intersection, but this intersection is also the midpoint of the two lines. Both rhombuses and squares have perpendicular bisecting angles, but trapezoids, rectangles, and parallelograms do not. Knowing these definitions can help tremendously in problems that involve both angles and diagonals.

Polygons with More Than Four Sides

A **pentagon** is a five-sided figure. A six-sided shape is a **hexagon**. A seven-sided figure is classified as a **heptagon**, and an eight-sided figure is called an **octagon**. An important characteristic is whether a polygon is regular or irregular. If it's **regular,** the side lengths and angle measurements are all equal. An **irregular** polygon has unequal side lengths and angle measurements. Mathematical problems involving polygons with more than four sides usually involve side length and angle measurements. The sum of all internal angles in a polygon equals $180(n-2)$ degrees, where n is the number of sides. Therefore, the total of all internal angles in a pentagon is 540 degrees because there are five sides so $180(5-2) = 540$ degrees. Unfortunately, area formulas don't exist for polygons with more than four sides. However, their shapes can be split up into triangles, and the formula for area of a triangle can be applied and totaled to obtain the area for the entire figure.

Probability and Statistics

Data Collection from Measurements on a Single Variable

Representing Data

Most statistics involve collecting a large amount of data, analyzing it, and then making decisions based on previously known information. These decisions also can be measured through additional data collection and then analyzed. Therefore, the cycle can repeat itself over and over. Representing the data visually is a large part of the process, and many plots on the real number line exist that allow this to be done. For example, a **dot plot** uses dots to represent data points above the number line. Also, a **histogram** represents a data set as a collection of rectangles, which illustrate the frequency distribution of the data. Finally, a **box plot** (also known as a **box and whisker plot**) plots a data set on the number line by segmenting the distribution into four quartiles that are divided equally in half by the median. Here's an example of a box plot, a histogram, and a dot plot for the same data set:

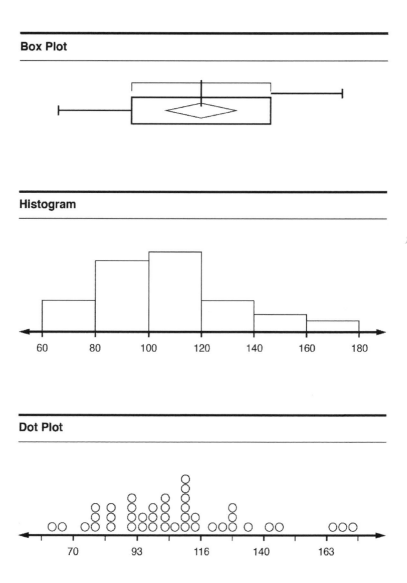

Comparing Data

Comparing data sets within statistics can mean many things. The first way to compare data sets is by looking at the center and spread of each set. The **center** of a data set can mean two things: median or mean. The **median** is the value that's halfway into each data set, and it splits the data into two intervals. The **mean** is the average value of the data within a set. It's calculated by adding up all of the data in the set and dividing the total by the number of data points. **Outliers** can significantly impact the mean. Additionally, two completely different data sets can have the same mean. For example, a data set with values ranging from 0 to 100 and a data set with values ranging from 44 to 56 can both have means of 50. The first data set has a much wider range, which is known as the **spread** of the data. This measures how varied the data is within each set. Spread can be defined further as either interquartile range or standard deviation. The **interquartile range (IQR)** is the range of the middle 50 percent of the data set. This range can be seen in the large rectangle on a box plot. The **standard deviation (σ)** quantifies the amount of variation with respect to the mean. A lower standard deviation shows that the data set doesn't differ greatly from the mean. A larger standard deviation shows that the data set is spread out farther from the mean. The formula for standard deviation is:

$$\sigma = \sqrt{\frac{\sum(x - \bar{x})^2}{n - 1}}$$

x is each value in the data set, $\bar{x}$ is the mean, and n is the total number of data points in the set.

Interpreting Data

The shape of a data set is another way to compare two or more sets of data. If a data set isn't symmetric around its mean, it's said to be **skewed**. If the tail to the left of the mean is longer, it's said to be **skewed to the left**. In this case, the mean is less than the median. Conversely, if the tail to the right of the mean is longer, it's said to be **skewed to the right** and the mean is greater than the median. When classifying a data set according to its shape, its overall **skewness** is being discussed. If the mean and median are equal, the data set isn't skewed; it is **symmetric**.

An **outlier** is a data point that lies a great distance away from the majority of the data set. It also can be labelled as an **extreme value**. Technically, an outlier is any value that falls 1.5 times the IQR above the upper quartile or 1.5 times the IQR below the lower quartile. The effect of outliers in the data set is seen visually because they affect the mean. If there's a large difference between the mean and mode, outliers are the cause. The mean shows bias towards the outlying values. However, the median won't be affected as greatly by outliers.

Normal Distribution

A **normal distribution** of data follows the shape of a bell curve and the data set's median, mean, and mode are equal. Therefore, 50 percent of its values are less than the mean and 50 percent are greater than the mean. Data sets that follow this shape can be generalized using normal distributions. Normal distributions are described as **frequency distributions** in which the data set is plotted as percentages rather than true data points. A **relative frequency distribution** is one where the y-axis is between zero and 1, which is the same as 0% to 100%. Within a standard deviation, 68 percent of the values are within 1 standard deviation of the mean, 95 percent of the values are within 2 standard deviations of the mean, and 99.7 percent of the values are within 3 standard deviations of the mean. The number of standard deviations that a data point falls from the mean is called the **z-score.** The formula for the z-score is $Z = \frac{x - \mu}{\sigma}$, where μ is the mean, σ is the standard deviation, and x is the data point. This formula is used to fit any data set that resembles a normal distribution to a standard normal distribution, in a process known as **standardizing**. Here is a normal distribution with labelled z-scores:

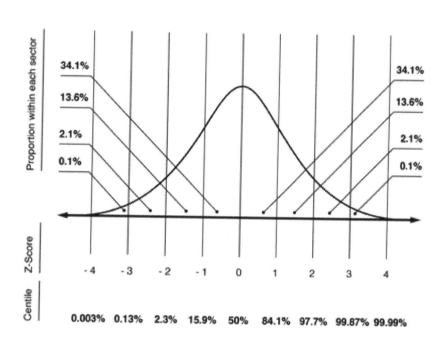

Normal Distribution with Labelled Z-Scores

Population percentages can be estimated using normal distributions. For example, the probability that a data point will be less than the mean, or that the z-score will be less than 0, is 50%. Similarly, the probability that a data point will be within 1 standard deviation of the mean, or that the z-score will be between -1 and 1, is about 68.2%. When using a table, the left column states how many standard deviations (to one decimal place) away from the mean the point is, and the row heading states the second decimal place. The entries in the table corresponding to each column and row give the probability, which is equal to the area.

Areas Under the Curve
The area under the curve of a standard normal distribution is equal to 1. Areas under the curve can be estimated using the z-score and a table. The area is equal to the probability that a data point lies in that region in decimal form. For example, the area under the curve from $z = -1$ to $z = 1$ is 0.682.

Data Collection from Measurements on Two Variables

Two-Way Frequency Tables
Data that isn't described using numbers is known as **categorical data**. For example, age is numerical data, but hair color is categorical data. Categorical data is summarized using two-way frequency tables. A **two-way frequency table** counts the relationship between two sets of categorical data. There are

rows and columns for each category, and each cell represents frequency information that shows the actual data count between each combination.

For example, the graphic on the left-side below is a two-way frequency table showing the number of girls and boys taking language classes in school. Entries in the middle of the table are known as the **joint frequencies**. For example, the number of girls taking French class is 12, which is a joint frequency. The totals are the **marginal frequencies**. For example, the total number of boys is 20, which is a marginal frequency. If the frequencies are changed into percentages based on totals, the table is known as a **two-way relative frequency table**. Percentages can be calculated using the table total, the row totals, or the column totals. Here's the process of obtaining the two-way relative frequency table using the table total:

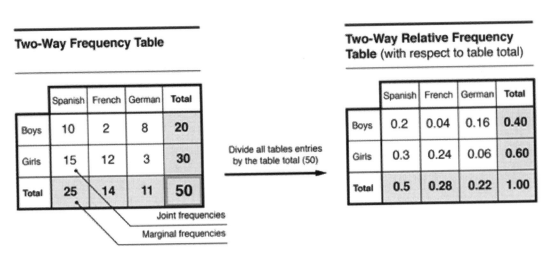

The middle entries are known as **joint probabilities** and the totals are **marginal probabilities.** In this data set, it appears that more girls than boys take Spanish class. However, that might not be the case because more girls than boys were surveyed and the results might be misleading. To avoid such errors, **conditional relative frequencies** are used. The relative frequencies are calculated based on a row or column.

Here are the conditional relative frequencies using column totals:

Two-Way Frequency Table

	Spanish	French	German	Total
Boys	10	2	8	20
Girls	15	12	3	30
Total	25	14	11	50

Divide each column entry by that column's total

Two-Way Relative Frequency Table (with respect to table total)

	Spanish	French	German	Total
Boys	0.4	0.14	0.73	0.4
Girls	0.6	0.86	0.27	0.6
Total	1.00	1.00	1.00	1.00

Data Conclusions

Two-way frequency tables can help in making many conclusions about the data. If either the row or column of conditional relative frequencies differs between each row or column of the table, then an association exists between the two categories. For example, in the above tables, the majority of boys are taking German while the majority of girls are taking French. If the frequencies are equal across the rows, there is no association and the variables are labelled as independent. It's important to note that the association does exist in the above scenario, though these results may not occur the next semester when students are surveyed.

Plotting Variables

A **scatter plot** is a way to visually represent the relationship between two variables. Each variable has its own axis, and usually the **independent** variable is plotted on the horizontal axis while the **dependent** variable is plotted on the vertical axis. Data points are plotted in a process that's similar to how ordered pairs are plotted on an *xy*-plane. Once all points from the data set are plotted, the scatter plot is finished. Below is an example of a scatter plot that's plotting the quality and price of an item. Note that price is the independent variable and quality is the dependent variable:

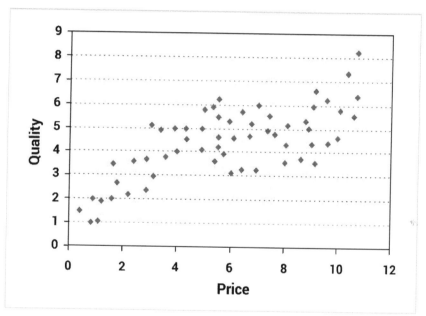

In this example, the quality of the item increases as the price increases.

Creating and Interpreting Linear Regression Models

Linear Regression

Regression lines are a way to calculate a relationship between the independent variable and the dependent variable. A straight line means that there's a linear **trend** in the data. Technology can be used to find the equation of this line (e.g., a graphing calculator or Microsoft Excel®). In either case, all of the data points are entered, and a line is "fit" that best represents the shape of the data. Other functions used to model data sets include quadratic and exponential models.

Estimating Data Points

Regression lines can be used to estimate data points not already given. For example, if an equation of a line is found that fit the temperature and beach visitor data set, its input is the average daily temperature and its output is the projected number of visitors. Thus, the number of beach visitors on a

100-degree day can be estimated. The output is a data point on the regression line, and the number of daily visitors is expected to be greater than on a 96-degree day because the regression line has a positive slope.

Plotting and Analyzing Residuals

Once the function is found that fits the data, its accuracy can be calculated. Therefore, how well the line fits the data can be determined. The difference between the actual dependent variable from the data set and the estimated value located on the regression line is known as a **residual.** Therefore, the residual is known as the predicted value $\hat{y}$ minus the actual value y. A residual is calculated for each data point and can be plotted on the scatterplot. If all the residuals appear to be approximately the same distance from the regression line, the line is a good fit. If the residuals seem to differ greatly across the board, the line isn't a good fit.

Interpreting the Regression Line

The formula for a regression line is $y = mx + b$, where m is the slope and b is the y-intercept. Both the slope and y-intercept are found in the **Method of Least Squares**, which is the process of finding the equation of the line through minimizing residuals. The slope represents the rate of change in y as x gets larger. Therefore, because y is the dependent variable, the slope actually provides the predicted values given the independent variable. The y-intercept is the predicted value for when the independent variable equals zero. In the temperature example, the y-intercept is the expected number of beach visitors for a very cold average daily temperature of zero degrees.

Correlation Coefficient

The **correlation coefficient (r)** measures the association between two variables. Its value is between -1 and 1, where -1 represents a perfect negative linear relationship, 0 represents no relationship, and 1 represents a perfect positive linear relationship. A **negative linear relationship** means that as x values increase, y values decrease. A **positive linear relationship** means that as x values increase, y values increase. The formula for computing the correlation coefficient is:

$$r = \frac{n(\sum xy) - (\sum x)(\sum y)}{\sqrt{n(\sum x^2) - (\sum x)^2}\sqrt{n(\sum y^2) - (\Sigma y)^2}}$$

n is the number of data points

Both Microsoft Excel® and a graphing calculator can evaluate this easily once the data points are entered. A correlation greater than 0.8 or less than -0.8 is classified as "strong" while a correlation between -0.5 and 0.5 is classified as "weak."

Correlation Versus Causation

Correlation and causation have two different meanings. If two values are **correlated**, there is an association between them. However, correlation doesn't necessarily mean that one variable causes the other. **Causation** (or "cause and effect") occurs when one variable causes the other. Average daily temperature and number of beachgoers are correlated and have causation. If the temperature increases, the change in weather causes more people to go to the beach. However, alcoholism and smoking are correlated but don't have causation. The more someone drinks the more likely they are to smoke, but drinking alcohol doesn't cause someone to smoke.

Statistical Processes

Samples and Populations

Statistics involves making decisions and predictions about larger data sets based on smaller data sets. Basically, the information from one part or subset can help predict what happens in the entire data set or population at large. The entire process involves guessing, and the predictions and decisions may not be 100 percent correct all of the time; however, there is some truth to these predictions, and the decisions do have mathematical support. The smaller data set is called a **sample** and the larger data set (in which the decision is being made) is called a **population**. A **random sample** is used as the sample, which is an unbiased collection of data points that represents the population as well as it can. There are many methods of forming a random sample, and all adhere to the fact that every potential data point has a predetermined probability of being chosen.

Goodness of Fit

Goodness of fit tests show how well a statistical model fits a given data set. They allow the differences between the observed and expected quantities to be summarized to determine if the model is consistent with the results. The **Chi-Squared Goodness of Fit Test** (or **Chi-Squared Test** for short) is used with one categorical variable from one population, and it concludes whether or not the sample data is consistent with a hypothesized distribution. Chi-Squared is evaluated using the following formula:

$$\chi^2 = \sum \frac{(O - E)^2}{E}$$

where O is the observed frequency value and E is the expected frequency value. Also, the **degree of freedom** must be calculated, which is the number of categories in the data set minus one. Then a Chi-Squared table is used to test the data. The **degree of freedom value** and a **significance value**, such as 0.05, are located on the table. The corresponding entry represents a critical value.

If the calculated χ^2 is greater than the critical value, the data set does not work with the statistical model. If the calculated χ^2 is less than the critical value, the statistical model can be used.

Making Inferences and Justifying Conclusions from Samples, Experiments, and Observational Studies

Data Gathering Techniques

The three most common types of data gathering techniques are sample surveys, experiments, and observational studies. **Sample surveys** involve collecting data from a random sample of people from a desired population. The measurement of the variable is only performed on this set of people. To have accurate data, the sampling must be unbiased and random. For example, surveying students in an advanced calculus class on how much they enjoy math classes is not a useful sample if the population should be all college students based on the research question. An **experiment** is the method in which a hypothesis is tested using a trial-and-error process. A cause and the effect of that cause are measured, and the hypothesis is accepted or rejected. Experiments are usually completed in a controlled environment where the results of a control population are compared to the results of a test population. The groups are selected using a randomization process in which each group has a representative mix of the population being tested. Finally, an **observational study** is similar to an experiment. However, this design is used when there cannot be a designed control and test population because of circumstances (e.g., lack of funding or unrealistic expectations). Instead, existing control and test populations must be used, so this method has a lack of randomization.

Population Mean and Proportion

Both the population mean and proportion can be calculated using data from a sample. The **population mean (μ)** is the average value of the parameter for the entire population. Due to size constraints, finding the exact value of μ is impossible, so the mean of the sample population is used as an estimate instead. The larger the sample size, the closer the sample mean gets to the population mean. An alternative to finding μ is to find the *proportion* of the population, which is the part of the population with the given characteristic. The proportion can be expressed as a decimal, a fraction, or a percentage, and can be given as a single value or a range of values. Because the population mean and proportion are both estimates, there's a *margin of error*, which is the difference between the actual value and the expected value.

T-Tests

A *randomized experiment* is used to compare two treatments by using statistics involving a *t-test*, which tests whether two data sets are significantly different from one another. To use a t-test, the test statistic must follow a normal distribution. The first step of the test involves calculating the *t* value, which is given as $t = \frac{\bar{x}_1 - \bar{x}_2}{s_{\bar{x}_1 - \bar{x}_2}}$, where $\bar{x}_1$ and $\bar{x}_2$ are the averages of the two samples. Also,

$$s_{\bar{x}_1 - \bar{x}_2} = \sqrt{\frac{s_1^2}{n_1} + \frac{s_2^2}{n_2}}$$

where s_1 and s_2 are the standard deviations of each sample and n_1 and n_2 are their respective sample sizes. The *degrees of freedom* for two samples are calculated as the following (rounded to the lowest whole number).

$$df = \frac{(n_1 - 1) + (n_2 - 1)}{2}$$

Also, a significance level α must be chosen, where a typical value is $\alpha = 0.05$. Once everything is compiled, the decision is made to use either a *one-tailed test* or a *two-tailed test*. If there's an assumed difference between the two treatments, a one-tailed test is used. If no difference is assumed, a two-tailed test is used.

Analyzing Test Results

Once the type of test is determined, the t-value, significance level, and degrees of freedom are applied to the published table showing the *t* distribution. The row is associated with degrees of freedom and each column corresponds to the probability. The t-value can be exactly equal to one entry or lie between two entries in a row. For example, consider a t-value of 1.7 with degrees of freedom equal to 30. This *test statistic* falls between the *p* values of 0.05 and 0.025. For a one-tailed test, the corresponding *p* value lies between 0.05 and 0.025. For a two-tailed test, the *p* values need to be doubled so the corresponding *p* value falls between 0.1 and 0.05. Once the probability is known, this range is compared to α. If $p < \alpha$, the hypothesis is rejected. If $p > \alpha$, the hypothesis isn't rejected. In a two-tailed test, this scenario means the hypothesis is accepted that there's no difference in the two treatments. In a one-tailed test, the hypothesis is accepted, indicating that there's a difference in the two treatments.

Evaluating Completed Tests

In addition to applying statistical techniques to actual testing, evaluating completed tests is another important aspect of statistics. Reports can be read that already have conclusions, and the process can be evaluated using learned concepts. For example, deciding if a sample being used is appropriate. Other things that can be evaluated include determining if the samples are randomized or the results are significant. Once statistical concepts are understood, the knowledge can be applied to many applications.

Independence and Conditional Probability

Sample Subsets

A sample can be broken up into subsets that are smaller parts of the whole. For example, consider a sample population of females. The sample can be divided into smaller subsets based on the characteristics of each female. There can be a group of females with brown hair and a group of females that wear glasses. There also can be a group of females that have brown hair *and* wear glasses. This "and" relates to the *intersection* of the two separate groups of brunettes and those with glasses. Every female in that intersection group has both characteristics. Similarly, there also can be a group of females that either have brown hair *or* wear glasses. The "or" relates to the union of the two separate groups of brunettes and glasses. Every female in this group has at least one of the characteristics. Finally, the group of females who do not wear glasses can be discussed. This "not" relates to the *complement* of the glass-wearing group. No one in the complement has glasses. *Venn diagrams* are useful in highlighting these ideas. When discussing statistical experiments, this idea can also relate to events instead of characteristics.

Verifying Independent Events

Two events aren't always independent. For examples, females with glasses and brown hair aren't independent characteristics. There definitely can be overlap because females with brown hair can wear glasses. Also, two events that exist at the same time don't have to have a relationship. For example, even if all females in a given sample are wearing glasses, the characteristics aren't related. In this case, the probability of a brunette wearing glasses is equal to the probability of a female being a brunette multiplied by the probability of a female wearing glasses. This mathematical test of $P(A \cap B) = P(A)P(B)$ verifies that two events are independent.

Conditional Probability

Conditional probability is the probability that event A will happen given that event B has already occurred. An example of this is calculating the probability that a person will eat dessert once they have eaten dinner. This is different than calculating the probability of a person just eating dessert.

The formula for the conditional probability of event A occurring given B is:

$$P(A|B) = \frac{P \ (A \text{ and } B)}{P(B)}$$

It's defined to be the probability of both A and B occurring divided by the probability of event B occurring. If A and B are independent, then the probability of both A and B occurring is equal to $P(A)P(B)$, so $P(A|B)$ reduces to just $P(A)$. This means that A and B have no relationship, and the probability of A occurring is the same as the conditional probability of A occurring given B. Similarly:

$$P(B|A) = \frac{P \ (B \text{ and } A)}{P(A)} = P(B)$$

(if A and B are independent)

<u>Independent Versus Related Events</u>
To summarize, conditional probability is the probability that an event occurs given that another event has happened. If the two events are related, the probability that the second event will occur changes if the other event has happened. However, if the two events aren't related and are therefore independent, the first event to occur won't impact the probability of the second event occurring.

Computing Probabilities of Simple Events, Probabilities of Compound Events, and Conditional Probabilities

<u>Simple and Compound Events</u>
A *simple event* consists of only one outcome. The most popular simple event is flipping a coin, which results in either heads or tails. A *compound event* results in more than one outcome and consists of more than one simple event. An example of a compound event is flipping a coin while tossing a die. The result is either heads or tails on the coin and a number from one to six on the die. The probability of a simple event is calculated by dividing the number of possible outcomes by the total number of outcomes. Therefore, the probability of obtaining heads on a coin is $\frac{1}{2}$, and the probability of rolling a 6 on a die is $\frac{1}{6}$. The probability of compound events is calculated using the basic idea of the probability of simple events. If the two events are independent, the probability of one outcome is equal to the product of the probabilities of each simple event. For example, the probability of obtaining heads on a coin and rolling a 6 is equal to $\frac{1}{2} \times \frac{1}{6} = \frac{1}{12}$. The probability of either A or B occurring is equal to the sum of the probabilities minus the probability that both A and B will occur. Therefore, the probability of obtaining either heads on a coin or rolling a 6 on a die is:

$$^1/_2 + ^1/_6 - ^1/_{12} = ^7/_{12}$$

The two events aren't mutually exclusive because they can happen at the same time. If two events are mutually exclusive, and the probability of both events occurring at the same time is zero, the probability of event A or B occurring equals the sum of both probabilities. An example of calculating the probability of two mutually exclusive events is determining the probability of pulling a king or a queen from a deck of cards. The two events cannot occur at the same time.

<u>Measuring Probabilities with Two-Way Frequency Tables</u>
When measuring event probabilities, two-way frequency tables can be used to report the raw data and then used to calculate probabilities. If the frequency tables are translated into relative frequency tables, the probabilities presented in the table can be plugged directly into the formulas for conditional probabilities. By plugging in the correct frequencies, the data from the table can be used to determine if events are independent or dependent.

<u>Differing Probabilities</u>
The probability that event A occurs differs from the probability that event A occurs given B. When working within a given model, it's important to note the difference. $P(A|B)$ is determined using the formula $P(A|B) = \frac{P(A \text{ and } B)}{P(B)}$ and represents the total number of A's outcomes left that could occur after B occurs. $P(A)$ can be calculated without any regard for B. For example, the probability of a student finding a parking spot on a busy campus is different once class is in session.

<u>The Addition Rule</u>
The probability of event A or B occurring isn't equal to the sum of each individual probability. The probability that both events can occur at the same time must be subtracted from this total. This idea is

shown in the *addition rule*: $P(A \text{ or } B) = P(A) + P(B) - P(A \text{ and } B)$. The addition rule is another way to determine the probability of compound events that aren't mutually exclusive. If the events are mutually exclusive, the probability of both A and B occurring at the same time is 0.

Uniform and Non-Uniform Probability Models

A *uniform probability model* is one where each outcome has an equal chance of occurring, such as the probabilities of rolling each side of a die. A *non-uniform probability model* is one where each outcome has an unequal chance of occurring. In a uniform probability model, the conditional probability formulas for $P(B|A)$ and $P(A|B)$ can be multiplied by their respective denominators to obtain two formulas for $P(A \text{ and } B)$. Therefore, the multiplication rule is derived as:

$$P(A \text{ and } B) = P(A)P(B|A) = P(B)P(A|B)$$

In a model, if the probability of either individual event is known and the corresponding conditional probability is known, the multiplication rule allows the probability of the joint occurrence of A and B to be calculated.

Binomial Experiments

In statistics, a **binomial experiment** is an experiment that has the following properties. The experiment consists of *n* repeated trial that can each have only one of two outcomes. It can be either a success or a failure. The probability of success, *p*, is the same in every trial. Each trial is also independent of all other trials. An example of a binomial experiment is rolling a die 10 times with the goal of rolling a 5. Rolling a 5 is a success while any other value is a failure. In this experiment, the probability of rolling a 5 is $\frac{1}{6}$. In any binomial experiment, x is the number of resulting successes, n is the number of trials, p is the probability of success in each trial, and $q = 1 - p$ is the probability of failure within each trial. The probability of obtaining x successes within n trials is:

$$P(X = x) = \frac{n!}{x!\,(n - x)!} p^x (1 - p)^{n-x}$$

With the following being the *binomial coefficient*:

$$\binom{n}{x} = \frac{n!}{x!\,(n - x)!}$$

Within this calculation, $n!$ is n factorial that's defined as:

$$n \cdot (n - 1) \cdot (n - 2) \dots 1$$

Let's look at the probability of obtaining 2 rolls of a 5 out of the 10 rolls.

Start with $P(X = 2)$, where 2 is the number of successes. Then fill in the rest of the formula with what is known, *n*=10, *x*=2, *p*=1/6, *q*=5/6:

$$P(X = 2) = \left(\frac{10!}{2!\,(10 - 2)!}\right)\left(\frac{1}{6}\right)^2\left(1 - \frac{1}{6}\right)^{10-2}$$

Which simplifies to:

$$P(X = 2) = \left(\frac{10!}{2!\,8!}\right)\left(\frac{1}{6}\right)^2\left(\frac{5}{6}\right)^8$$

Then solve to get:

$$P(X = 2) = \left(\frac{3628800}{80640}\right)(.0277)(.2325) = .2898$$

Making Informed Decisions Using Probabilities and Expected Values

Graphical Displays
Graphical displays are used to visually represent probability distributions in statistical experiments. Specific displays representing probability distributions illustrate the probability of each event. Histograms are typically used to represent probability distributions, and the actual probability can be thought of as $P(x)$, where x is the independent variable.

Expected Value
The **expected value** of a random variable represents the mean value in either a large sample size or after a large number of trials. According to the law of large numbers, after a large number of trials, the actual mean (and that of the probability distribution) is approximately equal to the expected value. The expected value is a weighted average and is calculated as $E(X) = \sum x_i p_i$, where x_i represents the value of each outcome and p_i represents the probability of each outcome. If all probabilities are equal, the expected value is:

$$E(X) = \frac{x_1 + x_2 + \cdots + x_n}{n}$$

Expected value is often called the **mean of the random variable** and is also a measure of central tendency.

Calculating Theoretical Probabilities
Given a statistical experiment, a theoretical probability distribution can be calculated if the theoretical probabilities are known. The theoretical probabilities are plugged into the formula for both the binomial probability and the expected value. An example of this is any scenario involving rolls of a die or flips of a coin. The theoretical probabilities are known without any observed experiments. Another example of this is finding the theoretical probability distribution for the number of correct answers obtained by guessing a specific number of multiple choice questions on a class exam.

Determining Unknown Probabilities
Empirical data is defined as real data. If real data is known, approximations concerning samples and populations can be obtained by working backwards. This scenario is the case where theoretical probabilities are unknown, and experimental data must be used to make decisions. The sample data (including actual probabilities) must be plugged into the formulas for both binomial probability and the expected value. The actual probabilities are obtained using observation and can be seen in a probability distribution. An example of this scenario is determining a probability distribution for the number of televisions per household in the United States, and determining the expected number of televisions per household as well.

Weighing Outcomes
Calculating if it's worth it to play a game or make a decision is a critical part of probability theory. Expected values can be calculated in terms of payoff values, and deciding whether to make a decision or play a game can be done based on the actual expected value. Applying this theory to gambling and card games is fairly typical. The payoff values in these instances are the actual monetary totals.

Fairness

Fairness can be used when making decisions given different scenarios. For example, a game of chance can be deemed fair if every outcome has an equal probability of occurring. Also, a decision or choice can be labeled as fair if each possible option has an equal probability of being chosen. Using basic probability knowledge allows one to make decisions based on fairness. Fairness helps determine if an event's outcome is truly random and no bias is involved in the results. Random number generators are a good way to ensure fairness. An example of an event that isn't fair is the rolling of a weighted die.

Using Simulations to Construct Experimental Probability Distributions and to Make Informal Inferences

Simulations

Simulations are experiments where devices such as coins, cards, and dice are used to generate results representing real outcomes. The simulations are used to estimate probabilities that replace theoretical probabilities in cases where collecting empirical data or using theoretical probability models is improbable or even impossible. Simulation allows experiments to be performed that closely resemble a real scenario. An example of this is determining how likely it is for a family with four children to have all girls. A coin toss can simulate this probability distribution, where heads means having a girl and tails means having a boy. Dice also can be used to simulate this process and calculate a similar experimental probability distribution by assigning even-numbered rolls to girls and odd-numbered rolls to boys. Performing either simulation would be a much less expensive process than surveying a large number of families with four children.

Statistical Inferences

Statistical inference is defined as the process in which properties are deduced involving data analysis completed on a given probability distribution. If completed simulations can estimate theoretical probability distributions, that data can be used in the decision-making process. The process of defining hypotheses and deriving estimates can be applied to the data stemming from the simulation. The assumed population is a data set that's larger than the simulation's data set. Therefore, the sample data can be used to make informal inferences about the larger population without 100% certainty.

Probabilities Involving Finite Sample Spaces and Independent Trials

Fundamental Counting Principle

The **fundamental counting principle** states that if there are m possible ways for an event to occur, and n possible ways for a second event to occur, there are $m \times n$ possible ways for both events to occur. For example, there are two events that can occur after flipping a coin and six events that can occur after rolling a die, so there are $2 \times 6 = 12$ total possible event scenarios if both are done simultaneously. This principle can be used to find probabilities involving finite sample spaces and independent trials because it calculates the total number of possible outcomes. For this principle to work, the events must be independent of each other.

Discrete Mathematics

Sequences

A **sequence** is an enumerated set of numbers, and each term or member is defined by the number it represents within the sequence. It can be **recursively defined**, which means each term is defined using prior terms. Also, it can be **explicitly defined** using only the number it represents within the sequence.

The **Fibonacci numbers** are a famous recursively defined sequence where the first and second terms are equal to 1 and every other term is equal to the sum of the two previous terms. Therefore, the first six Fibonacci numbers are 1, 1, 2, 3, 5, and 8. An example of an explicitly defined sequence is one where the nth term is $f_n = 2n + 1$. The first six terms of this sequence are 3, 5, 7, 9, 11, and 13. Both types of sequences can be used to model situations in the same way that functions are used to model real-life applications.

Number Patterns

Given a sequence of numbers, a mathematical rule can be defined that represents the numbers if a pattern exists within the set. For example, consider the sequence of numbers 1, 4, 9, 16, 25, etc. This set of numbers represents the positive integers squared, and an explicitly defined sequence that represents this set is $f_n = n^2$. An important mathematical concept is recognizing patterns in sequences and translating the patterns into an explicit formula. Once the pattern is recognized and the formula is defined, the sequence can be extended easily. For example, the next three numbers in the sequence are 36, 49, and 64.

Predicting Values

In a similar sense, patterns can be used to make conjectures, predictions, and generalizations. If a pattern is recognized in a set of numbers, values can be predicted that aren't originally provided. For example, if an experiment results in the sequence of numbers 1, 4, 9, 16, and 25, where 1 represents the first trial, 2 represents the second trial, etc., one expects the tenth trial to result in a value of 100 because that value is equal to the square of the trial number.

Recursion

Recursively Defined Functions

Similar to recursively defined sequences, **recursively defined functions** are not explicitly defined in terms of a variable. A recursive function builds on itself and consists of a smaller argument, such as $f(0)$ or $f(1)$ and the actual definition of the function. For example, a recursively defined function is the following:

$$f(0) = 3$$

$$f(n) = f(n - 1) + 2n$$

Contrasting an explicitly defined function, a recursively defined function must be evaluated in order. The first five terms of this function are:

$$f(0) = 3, f(1) = 5$$

$$f(2) = 9, f(3) = 15$$

$$f(4) = 23$$

Some recursively defined functions have an explicit counterpart and, like sequences, they can be used to model real-life applications. The Fibonacci numbers can also be thought of as a recursively defined function if $f(n) = f_n$.

Closed-Form Functions

A **closed-form function** can be evaluated using a finite number of operations such as addition, subtraction, multiplication, and division. An example of a function that's not a closed-form function is one involving an infinite sum. For example, $y = \sum_{n=1}^{\infty} x$ isn't a closed-form function because it consists of a sum of infinitely many terms. Many recursively defined functions can be expressed as a closed-form expression. To convert to a closed-form expression, a formula must be found for the nth term. This means that the recursively defined sequence must be converted to its explicit formula.

Equivalence Relations

Relations and Binary Relations
A **relation** of set A to set B is a subset of $A \times B$, which is the product of A and B. $A \times B$ is defined as the set of all ordered pairs (a, b) where a is in A and b is in B. A **binary relation** R from A to B is a subset of $A \times B$. The domain of R is the set of all first elements of the ordered pairs, and the range of R is the set of all second elements of the ordered pairs. A relation R is **reflective** if, for every a in A, (a, a) is in R. R is not reflective if there is an a in A such that (a, a) is not in R. A relation R is **symmetric** if, whenever (a, b) is in R, (b, a) is also in R. A relation R is not symmetric if there exists an (a, b) in R where (b, a) is not in R. Finally, a relation R is **transitive** if there exist two ordered pairs (a, b) and (b, c) that are both in R and (a, c) is in R. R is not transitive if (a, b) and (b, c) are in R, but (a, c) is not in R. An example of a relation that is reflexive, symmetric, and transitive is R^2. This is the set of all ordered pairs where each element can be any real number.

Equivalence Relations
Once the terminology of a relation being reflexive, symmetric, and transitive is understood, it can be determined whether a relation is an equivalence relation. An **equivalence relation** is a relation R that's reflexive, symmetric, and transitive. Because R^2 is reflexive, symmetric, and transitive, it's also an equivalence relation.

Discrete and Continuous Representations

Discrete representations involve specific and defined values, while **continuous representations** involve values that can take on anything in a bounded or unbounded interval. In a discrete set of data, the number of values can be counted. However, a continuous set of data has an infinite number of data points. An example of a discrete data set can involve independent values represented as the integers from 1 to 10, and an example of a continuous data set can involve independent values represented as any real numbers from 1 to 10. The number of data points within the discrete set is 10, and there are an infinite number of data points in the continuous set. If a function maps each type of independent variable to a corresponding output value, the function using the discrete data set is labeled as a **discrete function**, and a function using the continuous data set is labeled as a **continuous function**. The graph of a discrete function involves only a finite number of data points, and the graph of a continuous function is a curve. Additional examples of discrete data are shoe sizes, zip codes, and natural numbers. Conversely, examples of continuous data are height, weight, and time.

Choosing the Representation to Use
Both discrete and continuous representations can be used to model various phenomena. A representation type is chosen based on the input data. If the data to be input into the function consists of a finite number of values, a discrete representation is used. However, if the data is infinite, a continuous representation is used. Most often, a discrete representation is used given the fact that experimental data is usually finite. A continuous representation can be estimated using various techniques (such as linear and quadratic regression) to transform the discrete representation into a continuous representation.

Logic

<u>Terminology of Logic</u>
Logic is defined as the science of reasoning, and the goal of studying logic is to distinguish correct reasoning from incorrect reasoning. Making **inferences** is the process of drawing conclusions from statements containing premises, data, or information. A **statement** or **proposition** is a sentence that's either true or false, but can't be both. An example of a true proposition is "Chicago is in Illinois." A proposition can also be **composite**, which means it's made up of sub-propositions and connectives such as *and*, *or*, or *not*. An example of a composition proposition is "Chicago is in Illinois and Buffalo is in New York." An **argument** is a collection of propositions in which one is defined as the **conclusion** and the rest are the **premises**. In a good argument, the premises correctly support the conclusion. However, the argument can be incorrect and the premises don't support the conclusion. Within the argument, the conclusion usually has an argument indicator such as the words *therefore*, *thus*, *consequently*, *in conclusion,* etc. A **deductive argument** is when the goal is to show that the conclusion must be true if the premises are true. An **inductive argument** is when the goal is to show that the conclusion is probably true if the premises are true. A **conditional statement** is a proposition in which a statement is implied given another statement. Here is an example of a conditional statement: "If the clock reads noon, the class is eating lunch."

<u>Symbols of Logic</u>
Logic has three basic **symbols** representing the three logical operations. The symbol for the **conjunction** "and" is $\wedge$, so the statement $p \wedge q$ is read "p and q." The symbol for the **disjunction** "or" is $\vee$, so the statement $p \vee q$ is read "p or q." Finally, the symbol for the negation "not" is $\neg$, so the statement $\neg p$ is read "not p." Conditional statements also have their own symbols. "If p then q" can be represented symbolically as $p \rightarrow q$, and the biconditional statement "p if and only if q" can be represented symbolically as $p \leftrightarrow q$.

<u>Statement Truths</u>
For the conjunction $p \wedge q$ to be true, both p and q must be true. Otherwise, $p \wedge q$ is false. For a disjunction $p \vee q$ to be true, at least one of the statements p and q has to be true; otherwise, $p \vee q$ is false. Therefore, the only case in which $p \vee q$ is false is if both p and q are false. For a negation $\neg p$ to be true, p must be false. If p is true, the negation is false. The conditional statement $p \rightarrow q$ is false only when p is true but q is false. All other cases represent the truth for the conditional statement. Therefore, if p is ever false, the statement is true. Finally, the biconditional statement is true only if p and q are both true or both false. The two other scenarios give a false statement.

Truth tables can help in determining the truth of statements and can be extended to propositions that are expressions constructed from logical variables p, q, etc. and multiple connectives. The truth of a proposition depends on the truth of the individual logical variables and the properties of the connectives. Here's a truth table for the proposition:

$$P(p, q, r) = (p \lor q) \land r$$

This is read "p or q and r":

p	q	r	$(p \lor q)$	$(p \lor q) \land r$
T	T	T	T	T
T	T	F	T	F
T	F	T	T	T
T	F	F	T	F
F	T	T	T	T
F	T	F	T	F
F	F	T	F	F
F	F	F	F	F

Statement Equivalence

Two propositions are labelled as **equivalent** if they have identical truth tables. Two truth tables must be constructed and compared to determine if they are equal. The *inverse* of a conditional statement $p \to q$ is constructed when each logical variable p and q is negated and written as $\neg p \to \neg q$. The **converse** of a conditional statement is constructed when the statement is flipped and written as $q \to p$. The **contrapositive** is constructed by negating each logical variable within the converse. The converse is therefore written as $\neg q \to \neg p$. Truth tables can show equivalence between these statements.

Counting Techniques

There are many **counting techniques** that can help solve problems involving counting possibilities. For example, the **Addition Principle** states that if there are m choices from Group 1 and n choices from Group 2, then $n + m$ is the total number of choices possible from Groups 1 and 2. For this to be true, the groups can't have any choices in common. The **Multiplication Principle** states that if Process 1 can be completed n ways and Process 2 can be completed m ways, the total number of ways to complete both Process 1 and Process 2 is $n \times m$. For this rule to be used, both processes must be independent of each other. Counting techniques also involve permutations. A **permutation** is an arrangement of elements in a set for which order must be considered. For example, if three letters from the alphabet are chosen, ABC and BAC are two different permutations. The multiplication rule can be used to determine the total number of possibilities.

If each letter can't be selected twice, the total number of possibilities is:

$$26 \times 25 \times 24 = 15{,}600$$

A formula can also be used to calculate this total. In general, the notation $P(n, r)$ represents the number of ways to arrange r objects from a set of n and, the formula is:

$$P(n, r) = \frac{n!}{(n - r)!}$$

In the previous example:

$$P(26, 3) = \frac{26!}{23!} = 15{,}600$$

Contrasting permutations, a **combination** is an arrangement of elements in which order doesn't matter. In this case, ABC and BAC are the same combination. In the previous scenario, there are six permutations that represent each single combination. Therefore, the total number of possible combinations is $15{,}600 \div 6 = 2{,}600$. In general, $C(n, r)$ represents the total number of combinations of n items selected r at a time where order doesn't matter, and the formula is:

$$C(n, r) = \frac{n!}{(n - r)!\ r!}$$

Therefore, the following relationship exists between permutations and combinations:

$$C(n, r) = \frac{P(n, r)}{n!} = \frac{P(n, r)}{P(r, r)}$$

Set theory

In set theory, a **set** is defined as a collection of objects. A set can be finite, infinite, or empty.

Venn diagrams can represent sets pictorially as circles where each circle represents an individual set. These diagrams can be used to represent sets and understand terminology within set theory. If two sets have shared elements, they overlap. The **intersection** of the two sets represents the shared region as shown below in #3. The **union** of two sets represents the entire region covered by each set, and includes both the shared and unshared regions as shown in #4. The **complement** of a set (with respect to a given universal set) represents all elements in the universal set that aren't in the original set. For example, the Venn diagram in #6 shows the complement of set A with respect to set U. Finally, the **difference** of two sets, A and B, is equal to the set of all elements in A that don't belong to B. The Venn diagram in #5 visually represents A − B as the shaded region.

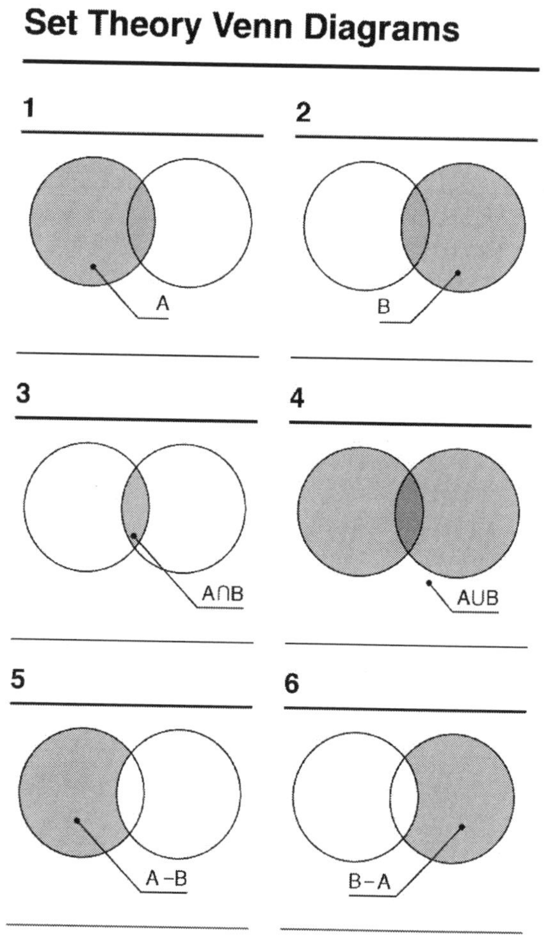

Set Theory Venn Diagrams

1

2

A

B

3

4

A∩B

A∪B

5

6

A−B

B−A

Practice Questions

1. A ball is drawn at random from a ball pit containing 8 red balls, 7 yellow balls, 6 green balls, and 5 purple balls. What's the probability that the ball drawn is yellow?

    a. $\dfrac{1}{26}$

    b. $\dfrac{19}{26}$

    c. $\dfrac{7}{26}$

    d. 1

2. Two cards are drawn from a shuffled deck of 52 cards. What's the probability that both cards are Kings if the first card isn't replaced after it's drawn and is a King?

    a. $\dfrac{1}{169}$

    b. $\dfrac{1}{221}$

    c. $\dfrac{1}{13}$

    d. $\dfrac{4}{13}$

3. What's the probability of rolling a 6 at least once in two rolls of a die?

    a. $\dfrac{1}{3}$

    b. $\dfrac{1}{36}$

    c. $\dfrac{1}{6}$

    d. $\dfrac{11}{36}$

4. Given the set $A = \{1, 2, 3, 4, 5, 6, 7, 8, 9, 10\}$ and $B = \{1, 2, 3, 4, 5\}$, what is $A - (A \cap B)$?

    a. $\{6, 7, 8, 9, 10\}$

    b. $\{1, 2, 3, 4, 5\}$

    c. $\{1, 2, 3, 4, 5, 6, 7, 8, 9, 10\}$

    d. $\emptyset$

5. Let p = "Alex is an engineering major," q = "Alex is not an English major," r = "Alex's sister is a history major," s = "Alex's sister has been to Germany," and t = "Alex's sister has been to Austria." Which of the following answers represents the statement "Alex is an engineering and English major, but his sister is a history major who hasn't been to either Germany or Austria."?

    a. $p \wedge \sim q \wedge (r \vee (\sim s \vee \sim t))$

    b. $p \wedge q \wedge r \vee (\sim s \wedge \sim t)$

    c. $p \wedge \sim q \wedge r \wedge (\sim s \vee \sim t)$

    d. $p \wedge q \wedge (r \vee (\sim s \wedge \sim t))$

6. For a group of 20 men, the median weight is 180 pounds and the range is 30 pounds. If each man gains 10 pounds, which of the following would be true?
    a. The median weight will increase, and the range will remain the same.
    b. The median weight and range will both remain the same.
    c. The median weight will stay the same, and the range will increase.
    d. The median weight and range will both increase.

7. For the following similar triangles, what are the values of x and y (rounded to one decimal place)?

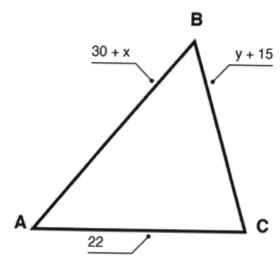 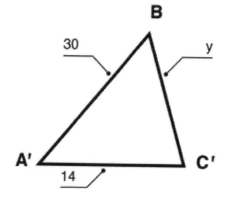

    a. $x = 16.5, y = 25.1$
    b. $x = 19.5, y = 24.1$
    c. $x = 17.1, y = 26.3$
    d. $x = 26.3, y = 17.1$

8. What are the center and radius of a circle with equation $4x^2 + 4y^2 - 16x - 24y + 51 = 0$?
    a. Center (3, 2) and radius ½
    b. Center (2, 3) and radius ½
    c. Center (3, 2) and radius ¼
    d. Center (2, 3) and radius ¼

9. If the ordered pair $(-3, -4)$ is reflected over the x-axis, what's the new ordered pair?
    a. $(-3, -4)$
    b. $(3, -4)$
    c. $(3, 4)$
    d. $(-3, 4)$

10. If the volume of a sphere is 288π cubic meters, what are the radius and surface area of the same sphere?
    a. Radius 6 meters and surface area 144π square meters
    b. Radius 36 meters and surface area 144π square meters
    c. Radius 6 meters and surface area 12π square meters
    d. Radius 36 meters and surface area 12π square meters

11. The triangle shown below is a right triangle. What's the value of x?

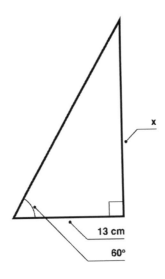

13 cm

60°

a. $x = 1.73$
b. $x = 0.57$
c. $x = 13$
d. $x = 22.49$

12. What's the midpoint of a line segment with endpoints $(-1, 2)$ and $(3, -6)$?
a. $(1, 2)$
b. $(1, 0)$
c. $(-1, 2)$
d. $(1, -2)$

13. A sample data set contains the following values: 1, 3, 5, 7. What's the standard deviation of the set?
a. 2.58
b. 4
c. 6.23
d. 1.1

14. Given the recursively defined sequence $a_1 = 9, a_n = a_{n-1} + 6$, which of the following is an explicit formula that represents the same sequence of numbers?
a. $a_n = 6(n-1) + 9$
b. $a_n = 6n + 9$
c. $a_n = 9n + 6$
d. $a_n = a_n + 6$

15. What is $\cos\frac{\pi}{8}$ evaluated exactly?
a. $\frac{\sqrt{2+\sqrt{3}}}{2}$
b. $\frac{\sqrt{2+\sqrt{2}}}{2}$
c. 0.9
d. 1

16. Given the following triangle, what's the length of the missing side? Round the answer to the nearest tenth.

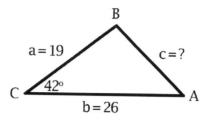

a. 17.0
b. 17.4
c. 18.0
d. 18.4

17. What are the coordinates of the focus of the parabola $y = -9x^2$?
   a. $(-3, 0)$
   b. $\left(-\frac{1}{36}, 0\right)$
   c. $(0, -3)$
   d. $\left(0, -\frac{1}{36}\right)$

18. How many possible two-number combinations are there for the numbers 1, 2, 3, 4, and 5 if each number can only be used once in any combination and order DOES matter?
   a. 120
   b. 60
   c. 20
   d. 10

19. The radius of a sphere is increasing at a rate of 6 cm/s. How fast is the volume increasing when the diameter is 40 cm? The formula for volume of a sphere with radius r is $V = \frac{4}{3}\pi r^3$.
   a. $9,600\pi$ cm/s
   b. 960π cm^3/s
   c. $9,600\pi$ cm^3/s
   d. 960π cm/s

20. A pair of dice is thrown, and the sum of the two scores is calculated. What's the expected value of the roll? Enter your answer in the answer box below.

Answer Explanations

1. C: The sample space is made up of $8 + 7 + 6 + 5 = 26$ balls. The probability of pulling each individual ball is $\frac{1}{26}$. Since there are 7 yellow balls, the probability of pulling a yellow ball is $\frac{7}{26}$.

2. B: For the first card drawn, the probability of a King being pulled is $\frac{4}{52}$. Since this card isn't replaced, if a King is drawn first, the probability of a King being drawn second is $\frac{3}{51}$. The probability of a King being drawn in both the first and second draw is the product of the two probabilities: $\frac{4}{52} \times \frac{3}{51} = \frac{12}{2,652}$ which, divided by 12, equals $\frac{1}{221}$.

3. D: The addition rule is necessary to determine the probability because a 6 can be rolled on either roll of the die. The rule used is $P(A \text{ or } B) = P(A) + P(B) - P(A \text{ and } B)$. The probability of a 6 being individually rolled is $\frac{1}{6}$ and the probability of a 6 being rolled twice is $\frac{1}{6} \cdot \frac{1}{6} = \frac{1}{36}$. Therefore, the probability that a 6 is rolled at least once is $\frac{1}{6} + \frac{1}{6} - \frac{1}{36} = \frac{11}{36}$.

4. A: $(A \cap B)$ is equal to the intersection of the two sets A and B, which is $\{1, 2, 3, 4, 5\}$. $A - (A \cap B)$ is equal to the elements of A that are *not* included in the set $(A \cap B)$. Therefore, $A - (A \cap B) = \{6, 7, 8, 9, 10\}$.

5. C: "Alex is an engineering and English major, but his sister is a history major who hasn't been to either Germany or Austria" can be rewritten as "p and not q and r and not s or not t." Using logical symbols, this is written as $p \wedge \sim q \wedge r \wedge (\sim s \vee \sim t)$.

6. A: If each man gains 10 pounds, every original data point will increase by 10 pounds. Therefore, the man with the original median will still have the median value, but that value will increase by 10. The smallest value and largest value will also increase by 10 and, therefore, the difference between the two won't change. The range does not change in value and, thus, remains the same.

7. C: Because the triangles are similar, the lengths of the corresponding sides are proportional. Therefore:

$$\frac{30 + x}{30} = \frac{22}{14} = \frac{y + 15}{y}$$

This results in the equation $14(30 + x) = 22 \cdot 30$ which, when solved, gives $x = 17.1$. The proportion also results in the equation $14(y + 15) = 22y$ which, when solved, gives $y = 26.3$.

8. B: The technique of completing the square must be used to change $4x^2 + 4y^2 - 16x - 24y + 51 = 0$ into the standard equation of a circle. First, the constant must be moved to the right-hand side of the equals sign, and each term must be divided by the coefficient of the x^2 term (which is 4). The x and y terms must be grouped together to obtain:

$$x^2 - 4x + y^2 - 6y = -\frac{51}{4}$$

Then, the process of completing the square must be completed for each variable. This gives:

$$(x^2 - 4x + 4) + (y^2 - 6y + 9) = -\frac{51}{4} + 4 + 9$$

The equation can be written as:

$$(x - 2)^2 + (y - 3)^2 = \frac{1}{4}$$

Therefore, the center of the circle is (2, 3) and the radius is:

$$\sqrt{\frac{1}{4}} = \frac{1}{2}$$

9. D: When an ordered pair is reflected over an axis, the sign of one of the coordinates must change. When it's reflected over the x-axis, the sign of the x coordinate must change. The y value remains the same. Therefore, the new ordered pair is $(-3, 4)$.

10. A: Because the volume of the given sphere is 288π cubic meters, this means $\frac{4}{3}\pi r^3 = 288\pi$. This equation is solved for r to obtain a radius of 6 meters. The formula for the surface area of a sphere is $4\pi r^2$, so if $r = 6$ in this formula, the surface area is 144π square meters.

11. D: SOHCAHTOA is used to find the missing side length. Because the angle and adjacent side are known, $\tan 60 = \frac{x}{13}$. Making sure to evaluate tangent with an argument in degrees, this equation gives:

$$x = 13 \tan 60 = 13 \cdot 1.73 = 22.49$$

12. D: The midpoint formula should be used.

$$M = \left(\frac{x_1 + x_2}{2}, \frac{y_1 + y_2}{2}\right) = \left(\frac{-1 + 3}{2}, \frac{2 + (-6)}{2}\right) = (1, -2)$$

13. A: First, the sample mean must be calculated:

$$\bar{x} = \frac{1}{4}(1 + 3 + 5 + 7) = 4$$

The standard deviation of the data set is $\sigma = \sqrt{\frac{\Sigma(x - \bar{x})^2}{n-1}}$, and $n = 4$ represents the number of data points. Therefore,

$$\sigma = \sqrt{\frac{1}{3}[(1 - 4)^2 + (3 - 4)^2 + (5 - 4)^2 + (7 - 4)^2]} = \sqrt{\frac{1}{3}(9 + 1 + 1 + 9)} = 2.58$$

14. A: An explicit formula is derived by evaluating a handful of terms in the recursively defined formula until a pattern is seen. In this example,

$$a_1 = 9, a_2 = a_1 + 6 = 9 + 6, a_3 = a_2 + 6 = 9 + 6 + 6, a_4 = a_3 + 6 = 9 + 6 + 6 + 6$$

The pattern is that $a_n = 9 + 6(n - 1)$.

15. B: The decimal approximation is not an exact answer. In order to obtain an exact answer, the half-angle formula must be used as follows:

$$\cos\frac{\pi}{8} = \sqrt{\frac{1+\cos\frac{\pi}{4}}{2}} = \sqrt{\frac{1+\frac{\sqrt{2}}{2}}{2}} = \frac{\sqrt{2+\sqrt{2}}}{2}$$

The positive value was selected because the angle is located in quadrant I where all coordinates are positive.

16. B: Because this isn't a right triangle, SOHCAHTOA can't be used. However, the law of cosines can be used. Therefore:

$$c^2 = a^2 + b^2 - 2ab\cos C = 19^2 + 26^2 - 2\cdot 19\cdot 26\cdot\cos 42° = 302.773$$

Taking the square root and rounding to the nearest tenth results in $c = 17.4$.

17. D: A parabola of the form $y = \frac{1}{4f}x^2$ has a focus $(0, f)$. Because $y = -9x^2$, set $-9 = \frac{1}{4f}$. Solving this equation for f results in $f = -\frac{1}{36}$. Therefore, the coordinates of the focus are $\left(0, -\frac{1}{36}\right)$.

18. C: Because order *does* matter, the total number of permutations needs to be computed. The following represents the number of ways that two objects can be arranged from a set of five:

$$P(5,2) = \frac{5!}{(5-2)!} = \frac{120}{6} = 20$$

19. C: This is a related rates problem. When diameter is 40 cm, the radius is 20 cm. Volume and the radius are changing with respect to time, so the formula for volume must be integrated implicitly with respect to *t*. Therefore, $\frac{dV}{dt} = 4\pi r^2\frac{dr}{dt}$. The problem gives that $\frac{dr}{dt} = 6\frac{\text{mm}}{\text{s}}$, and therefore $\frac{dV}{dt} = 4\pi(20)^2 6 = 9,600\pi$. The units are cm^3/s because the amount represents the rate at which the volume is changing.

20. 7: The expected value is equal to the total sum of each product of individual score and probability. There are 36 possible rolls. The probability of rolling a 2 is $\frac{1}{36}$. The probability of rolling a 3 is $\frac{2}{36}$. The probability of rolling a 4 is $\frac{3}{36}$. The probability of rolling a 5 is $\frac{4}{36}$. The probability of rolling a 6 is $\frac{5}{36}$. The probability of rolling a 7 is $\frac{6}{36}$. The probability of rolling an 8 is $\frac{5}{36}$. The probability of rolling a 9 is $\frac{4}{36}$. The probability of rolling a 10 is $\frac{3}{36}$. The probability of rolling an 11 is $\frac{2}{36}$. Finally, the probability of rolling a 12 is $\frac{1}{36}$.

Each possible outcome is multiplied by the probability of it occurring. Like this:

$$2 \times \frac{1}{36} = a$$

$$3 \times \frac{2}{36} = b$$

$$4 \times \frac{3}{36} = c$$

And so forth.

Then all of those results are added together:

$$a + b + c \dots = expected\ value$$

In this case, it equals 7.

Dear Praxis II Mathematics Test Taker,

We would like to start by thanking you for purchasing this study guide for your Praxis II Mathematics exam. We hope that we exceeded your expectations.

Our goal in creating this study guide was to cover all of the topics that you will see on the test. We also strove to make our practice questions as similar as possible to what you will encounter on test day. With that being said, if you found something that you feel was not up to your standards, please send us an email and let us know.

We would also like to let you know about other books in our catalog that may interest you.

Praxis II Elementary Education Test

This can be found on Amazon: amazon.com/dp/1628454326

Praxis II English Language Arts

amazon.com/dp/1628454105

Praxis II General Science

amazon.com/dp/1628454385

Praxis II Social Studies

amazon.com/dp/1628454210

Praxis Core Study Guide

amazon.com/dp/1628454946

We have study guides in a wide variety of fields. If the one you are looking for isn't listed above, then try searching for it on Amazon or send us an email.

Thanks Again and Happy Testing!
Product Development Team
info@studyguideteam.com

Interested in buying more than 10 copies of our product? Contact us about bulk discounts:

bulkorders@studyguideteam.com

FREE Test Taking Tips DVD Offer

To help us better serve you, we have developed a Test Taking Tips DVD that we would like to give you for FREE. **This DVD covers world-class test taking tips that you can use to be even more successful when you are taking your test.**

All that we ask is that you email us your feedback about your study guide. Please let us know what you thought about it – whether that is good, bad or indifferent.

To get your **FREE Test Taking Tips DVD**, email freedvd@studyguideteam.com with "FREE DVD" in the subject line and the following information in the body of the email:

   a. The title of your study guide.

   b. Your product rating on a scale of 1-5, with 5 being the highest rating.

   c. Your feedback about the study guide. What did you think of it?

   d. Your full name and shipping address to send your free DVD.

If you have any questions or concerns, please don't hesitate to contact us at freedvd@studyguideteam.com.

Thanks again!

Made in the USA
Middletown, DE
29 January 2019